Commercial Conflict Management and Dispute Resolution

Commerce is inherently complex and the sums of money involved can be astronomical, so it is no surprise that conflicts and disputes are all too common. There are numerous techniques designed to resolve these problems, and this book summarises the most important of these, as well as alternative dispute resolution methods. The reader seeking a deeper understanding of these procedures will also find clear explanations of the principles and methods for conflict management, such as negotiation, risk management, mediation and conciliation.

As well as outlining these different techniques, guidance on which approach is appropriate in common situations is also given, helping the reader apply what they have learned to the real world. The significance of cultural issues is explained, before the reader is presented with suggestions for how to take these into account. Throughout, the book is illustrated with case studies from many different sectors as diverse as Mumbai's dabbawalla, the First World War and Terminal 5 at London Heathrow Airport.

Written with undergraduate students in mind, this book also serves to give a neat and brief overview for professionals. Those studying or working in commerce generally, construction project management, construction management and construction law will find this to be an invaluable book.

Peter Fenn is a Senior Lecturer at the University of Manchester and a Mediator and Adjudicator on the President's Panel at the Chartered Institute of Arbitrators (CIArb) and a Mediator on the President's Panel at the Royal Institution of Chartered Surveyors (RICS). He is also a trustee of CIArb and elected member of the Dispute Resolution Faculty Board at the RICS.

Commercial Conflict Management and Dispute Resolution

Peter Fenn

First published 2012
by Spon Press
2 Park Square, Milton Park, Abingdon, Oxon OX14 4RN

Simultaneously published in the USA and Canada
by Spon Press
711 Third Avenue, New York, NY 10017

Spon Press is an imprint of the Taylor & Francis Group, an informa business

British Library Cataloguing in Publication Data
A catalogue record for this book is available from the British Library

Library of Congress Cataloging in Publication Data
Fenn, Peter (Peter F.)
Commercial conflict management and dispute resolution / Peter Fenn.
p. cm.
Includes bibliographical references and index.
1. Conflict management. 2. Dispute resolution (Law) 3. Conflict management--Case studies. I. Title.
HD42.F457 2011
658.4'053--dc23
2011039544

ISBN: 978-0-415-57826-4 (hbk)
ISBN: 978-0-415-57828-8 (pbk)
ISBN: 978-0-203-85221-7 (ebk)

Typeset in Sabon
by GreenGate Publishing Services, Tonbridge, Kent

Printed and bound in Great Britain by
TJ International Ltd, Padstow, Cornwall

Contents

Figures and tables

Figures

Tables

Preface

This is a book brought about by many things. The author is a university academic and a long-standing resolver of disputes; he is a pragmatist but sees the need for underlying theory. He is bewildered by the plethora of textbooks which investigate individual conflict management or dispute resolution techniques. This is a short book about everything.

There are lots of books about particular dispute resolution techniques (e.g. arbitration, adjudication or mediation) but there are few which summarise all the techniques. There are few books about conflict management and dispute resolution.

This book first proposes a structure by defining conflict and dispute, then details conflict management and dispute resolution.

This is a book which summarises all the dispute resolution techniques as a 'first-stop' reference and as a guide to which is appropriate. In addition, the book considers conflict management techniques.

There are many long books about detailed issues, particularly in dispute resolution: long books about little issues (in dispute resolution). Again, this is a short book about everything.

Many people's thoughts have gone into this book, any mistakes remain mine. Siobhan Curley made sense of my poor drafting and the final text is greatly improved by her suggestions. David Lowe helped with case studies and general advice.

1 Introduction

Pink ribbon

Each chapter in this book has a pink ribbon; the reason is explained a bit later. The pink ribbon here concerns the central terms to this book: commercial conflict and dispute raise problems of definition. This book differentiates between conflict management and dispute resolution. Frank Sander, a Harvard professor, wrote a famous paper more than 50 years ago and this still frames much of the thinking about dispute resolution.

Introduction

This is a book about commercial conflict management and dispute resolution. That means starting with some definitions: commercial; conflict management and dispute resolution. Commercial is easier and quicker than conflict management and dispute resolution, which will take some time in later sections and chapters, but here goes: Commercial: commercial law (sometimes known as business law) is a number of laws that every business must adhere to.

So Commercial deals with business and commerce – that's hardly illuminating and might be considered a circular definition. One way of dealing with the circularity problem is to say what it is not – in this book commercial conflict management and dispute resolution are not concerned with consumer disputes or family disputes. This is a fiendishly difficult area and you might want to consider the art of rhetoric, particularly antimetabole and chiasmus.

In rhetoric, which is the art and study of the use of language, to carry an argument as use by persuasive orators, two techniques are used: antimetabole and chiasmus.

Antimetabole is the repetition of words in successive clauses, but in transposed grammatical order, for example: 'When a dog bites a man that is not news, but when a man bites a dog that is news' (Charles Anderson Dana, 'What is News?' *The New York Sun*, 1882).

Chiasmus is a figure of speech in which two or more clauses are related to each other through a reversal of structures in order to make a larger point, for example: I'd rather have a bottle in front of me than a frontal lobotomy.

Commercial is very difficult to define but you know what it means: it's axiomatic. You don't know what it means? Well, commercial does not include family or consumer law, but there are disputes (aplenty) in family and consumer issues. This definition problem is not unique to commercial issues, or to legal issues, although much debate and literature exists in the law. If you want to test how difficult it is; try and review the definitions of law or the law.

The United Nations Commission on International Trade Law (UNCITRAL) produced a model law: the UNCITRAL Model Law on International Commercial Arbitration. This provides (at the Note to Article 1 (1)) that it should be given wide interpretation so as to cover matters arising from all relationships of a commercial nature whether contractual or not.

Want some more evidence that definition of commercial is difficult? The New York Convention on the Recognition and Enforcement of Foreign Arbitral Awards (1958) allows countries a reservation, the commercial Reservation. Where a country avails themselves of the commercial Reservation they need only apply the provisions of the convention to disputes regarded as commercial by their own internal system of laws. As of January 2011, UNCITRAL give 145 countries as ratifying the Convention and 47 having taken advantage of the commercial reservation (see Appendix A).

This book makes a fundamental differentiation between conflict management and dispute resolution. You might disagree with it, many do, but the book, for reasons explained in the next chapter, is framed by the differentiation.

> Conflict management: here, the emphasis is on the axiom that it must be in all parties' interests to avoid disputes by managing conflict in such a way that disputes do not arise. This is sometimes described as dispute avoidance.
>
> Dispute resolution: Notwithstanding the emphasis on the desire to avoid dispute, there must be occasions where the parties have legitimate disputes and that the techniques of dispute resolution are employed to bring about the conclusion or resolution of the dispute.

The pink ribbon

A long time ago in a galaxy far away, a friend working at a law firm gave me some papers to help with some research. While he was out of the room I found some pink ribbon which I used to parcel all the papers together to take out with me. The senior partner of the law firm saw me in reception and asked what I was doing; I told him that I was taking some papers away. 'Not wrapped in ribbon like that you are not.' It transpired that bundling papers in pink ribbon was something only counsel (a barrister) could do. I learned my lesson. There were pink ribbon counterparts in many things I did over the years and I began to use it as an example of specialist knowledge or language that captured some 'secret'. Each chapter in this book has a pink ribbon. This describes the secret that might be hidden in the chapter.

The references and the case law

It is impossible to write any kind of book without providing references; these are a trail that others can pick up. But sometimes the references can be intimidating and overpowering, so this book does not include a comprehensive literature review. There are many such books, please look elsewhere. Similarly, any book about legal issues, at least in the common law, must include a detailed citation of case law. This book does not include a comprehensive citation of case law. Again, there are many such books, please look elsewhere. Of course, there are some references to literature and to cases but this is not a reference book.

Conflict and dispute

Conflict and dispute are difficult words; they are often interchanged. Some theorists take conflict as the stronger term (i.e. dispute is all around but only occasionally does conflict break out) and some take completely the opposite view (i.e. conflict is all around but only occasionally does dispute break out). This book takes the position that conflict is necessary and inevitable but that disputes are to be avoided. The classification used here is then Conflict Management or Dispute Resolution. Conflict is good; but effective management can avoid disputes. Dispute is bad and disputes need resolving (normally by third party intervention).

Dispute processing

Conflict management or dispute resolution or dispute processing? Because not all disputes get resolved, and many do not get resolved satisfactorily at all, some commentators use the term dispute processing rather than dispute resolution.

This book differentiates between conflict management and dispute resolution (or dispute processing) but it does NOT use the term dispute processing.

The term dispute processing was coined by a famous academic who wrote a famous paper. That academic is Frank Sander from Harvard University. The famous paper (*Perspectives on justice in the future*)[1] was first presented at a famous conference, The Pound conference, and is widely acknowledged as a seminal text on dispute resolution.

An alternative classification

To avoid the confusion between conflict and dispute it is possible to avoid the use of the term conflict at all and call everything dispute. Instead of conflict management, dispute avoidance. The continuum becomes:

Dispute avoidance ———————————————— Dispute resolution

Dispute resolution might be split into consensual and adjudicative. Here are some examples:

Dispute avoidance – project definition, risk allocation, procurement and tendering procedures, partnering, relationship contracting, negotiation techniques, dispute resolution adviser, project management.

Consensual dispute resolution – negotiation, dispute resolution adviser, mediation, conciliation neutral fact finding, neutral expert evaluation, mini-trial/executive tribunal, dispute review board.

Adjudicative dispute resolution – adjudication, expert determination, reference to an expert, dispute adjudication board, arbitration, litigation.

Litigation

Litigation is undoubtedly dispute resolution, the ultimate dispute resolution. Beyond this there is no treatment of litigation, other than to use it as a comparator, in this book. Len Shackleton was a famous soccer player and in his classic autobiography he included a chapter entitled 'What the average director knows about football' and he left the page blank. A chapter entitled 'What the author knows about litigation' and a page left blank was not approved by the publisher; however the author is very definitely not in the camp that blames any or all failings on lawyers: 'The first thing we do, let's kill all the lawyers' (Shakespeare Henry VI, Part 2).

In fact, the only piece of advice offered in the entire book is: the first thing you should do is consult a lawyer.

2 Conflict management and dispute resolution

Pink ribbon

There does exist a theory of conflict, proposed by Karl Marx and developed by others. There is considerable interest in conflict and disputes from a psychological, through a sociological to a commercial perspective. It used to be thought that all conflict was a bad thing. Among the first to question this was Mary Parker Follett, who developed the concepts of *functional* and *dysfunctional* conflict. The generally accepted view now is that conflict and dispute are different. The difference however is less easily explained. The widely held view is that conflict, which is all around in Western dialectic, may develop into dispute (although some argue the complete opposite that dispute is all around and conflict is the stronger term).

This book takes the view that conflict is inevitable, and is an essential part of dynamic capitalism. Dispute may flow from the conflict. If you like: conflict is inevitable, dispute is not. Dispute may emerge from conflict but conflict does not emerge from dispute. Some people talk of functional conflict and dysfunctional conflict or dispute. Therefore, two things are required: conflict management and dispute resolution.

Conflict management is considered in the next chapter. The dispute resolution techniques are considered individually in later chapters, but here the four major dispute resolution techniques are stated as: mediation, arbitration, construction adjudication and litigation. These are compared and contrasted under the headings: formality, speed, flexibility and cost.

The cost of commercial conflict and dispute is not easily quantifiable, but one thing is clear – whatever the cost is, it is something that should be avoided.

Those believing the argument that conflict is inevitable have to consider the example of the dabbawalla of Mumbai, where conflict is almost unheard of and it is estimated that less than one in six million deliveries produce mistakes. A case study of the dabbawalla is made.

Introduction

Both professionals and academics are enormously interested in commercial conflict and disputes. The interest is mostly with the techniques used to resolve disputes; there is little by way of interest into conflict management or dispute avoidance. This chapter considers four areas:

1. An introduction to conflict theory.
2. A discussion of the difference between conflict and dispute.
3. An outline of the myriad of techniques used to resolve disputes. In the UK this reflects the Government's current approach and that of other interests which seek to make savings by optimising efficiency in dispute resolution.
4. The lack of evidence that is apparent in any discussion of commercial disputes. This chapter discusses the absence of an empirical base to the study of disputes. The UK construction industry is given as an example. The lack of an empirical base means that there has been little consideration of the issues of understanding, explanation or prediction of commercial disputes. A research agenda is proposed where an aetiological approach to commercial disputes is employed; this, it is proposed, may help develop a mature and sophisticated research base, which may help industry performance.

A case study is made of the dabbawalla of Mumbai where 99.99966 per cent of the products manufactured (services delivered in this case) are statistically expected to be free of defects. Or, if you like, one in six million deliveries produce a mistake.

Conflict theory

A theory of conflict does exist, it was founded by Karl Marx.[1] Marx expresses the theory in terms of a class struggle – the struggle between classes. Others, notably Max Weber, took the theory forward. A glance at the conflict literature shows there is a great diversity of conflict knowledge, from the everyday knowledge we all have to the sophisticated theoretical writings of psychologists and sociologists. The problem is how to present this broad range of knowledge in an understandable manner. One way of dealing with this is to consider that the theories apply to many different conflicts, or even that they apply to all conflicts. Sociological theories apply to commercial conflict. In addition the theories will be presented in a simple way. One famous definition of economics is that it is a study of the allocation of scarce resources which have alternative uses. Conflict theory might be expressed in a similar way: conflict is inevitable as organisations seek to redistribute scarce resources – a classic Marxist view.

Conflict and dispute: is there a difference between conflict and disputes?

Disputes are time consuming, expensive and unpleasant. They can destroy client/supplier relationships which have been painstakingly built up over long periods of time. Disputes can add substantially to the cost of a project, even making the project unsuccessful, unfeasible or negating any benefits. Disputes need to be avoided, and if they cannot be avoided then they should be resolved as efficiently as possible to manage the 'problem', negotiate a 'settlement', help 'preserve relationships' and maintain 'value for money'.

Many people would not recognise a distinct difference between the terms conflict and dispute. Certainly most people would not concern themselves with any definition. Academics, and others, would usually make definition their starting point. Definition provides structure and structure may allow explanation and understanding.

Conflict and dispute studies do form academic disciplines. Any attempt here to summarise the various strands of academic disciplines would be doomed to failure and debate on definition. In an attempt to avoid this, the following are some suggested areas of conflict and dispute studies:

- Peace and conflict studies – a social science.
- Conflict management as an organisation management science – part of management science.
- Conflict management and dispute resolution – the concern of this book.

Functional and dysfunctional conflict

Early theory marked all conflict as a bad thing that should be avoided. Among the first to question this was Mary Parker Follett,[2] who said that effective conflict management ought not to conceive conflict as a wasteful outbreak of incompatibilities, but a normal process whereby socially valuable differences register themselves for the enrichment for all concerned. Three methods were advanced for dealing with conflict:

- *Domination* – there is a victory of one side over the other (a win–lose situation).
- *Compromise* – each side gives up something in the process (a lose–lose situation).
- *Integration* – each side refocuses their efforts so that neither side loses anything and each in fact gains (a win–win situation).

Parker Follett recommended only integration and this issue of integration is returned to in negotiation (see Chapter 4) and in mediation (see Chapter 5).

Follett believed that domination should be avoided at all costs. Although application of this strategy requires little effort on the part of the parties and their agents, the long-term side effects can be devastating. Compromise carries with it the assumption that both parties will be happy because each will gain something, but each loses something as well and this, in turn, creates the potential for further conflict. Integration was favoured simply because, if both parties can become satisfied, there will remain no issue or problem – obviously an ideal situation not easily attained.

Win–lose is often overused as a strategy for solving conflicts. It assumes the use of mental or physical power to bring about compliance; a lose–lose approach will also leave no one entirely happy. Compromise, side payments and submission of the issue to a neutral third party, as in the arbitration procedure, constitute examples of this latter approach. The win–win approach (now becoming more popular, although still misunderstood) yields solutions satisfactory to all, in that each party to the conflict wins something and the conflict is therefore resolved constructively. It could be suggested that important conflicts tend to be best managed with positive-sum (win–win) strategies, while more trivial issues merit no more than zero-sum (win–lose/lose–lose) strategies, with most situations calling for contingency or mixed modes (no win–no lose). The concepts of integration and interest-based approaches have considerable influence in negotiation, mediation and Game Theory (see Chapter 10).

A further distinction between conflict and dispute that is particularly useful is the one which distinguishes the two based on time and issues in contention.[3] Disputes, this suggests, are short-term disagreements that are relatively easy to resolve. Long-term, deep-rooted problems that involve seemingly non-negotiable issues and are resistant to resolution are referred to as conflicts. Though both types of disagreement can occur independently of one another, they can also be connected. In fact, one way to think about the difference between them is that short-term disputes might exist within a larger, longer conflict. A similar concept would be the notion of battles, which occur within the broader context of a war. Other theorists talk of strategy and tactics; tactics win the battle but strategy wins the war.

From this analysis of conflict and dispute it can be argued that conflict is necessary and inevitable, but that disputes are to be avoided. The school of Western thought maintains that conflict (but not dispute) is inevitable.[4] Conflict is part of Western societies and idioms; there is a Western dialectic argument idiom, to use the academic jargon. Conflict is part of dynamic capitalism and an integral part of commercialism, conflict might be seen as the functional and necessary part. Dispute, on the other hand, only develops when conflict is not (or cannot be) managed. Dispute, therefore, is the unnecessary or dysfunctional element. Logically, then, there are two areas for consideration:

1 *Conflict Management* – here the emphasis is on the axiom that it must be in all parties' interests to avoid disputes by managing conflict in such a way that disputes do not arise (sometimes described as dispute avoidance).

2 *Dispute Resolution* – notwithstanding the emphasis on the desire to avoid dispute, there must be occasions where the parties have legitimate disputes and that the techniques of dispute resolution are employed to bring about the conclusion or resolution of the dispute.

The distinction between conflict and dispute is shown diagrammatically in Figure 2.1. It might be argued that at the dispute end of the continuum lies other action (e.g. violence). Hopefully we will not have to consider this option.

Disputes on projects, or contracts, are more than unpleasant, they divert valuable resources from the overall aim, which must be: completed on time, on budget and to the quality specified. In addition, they generally cost money, take time and can destroy relationships, which may have taken years to develop.

The legal issue of conflict or dispute: legal concerns

Although the esoteric discussion earlier that conflict and dispute is valuable, is there any pragmatic real-world issue in the distinction? The legal point is discussed as the difference between behavioural conflict and justiciable dispute.[5] The question as to whether or not a dispute exists is highly

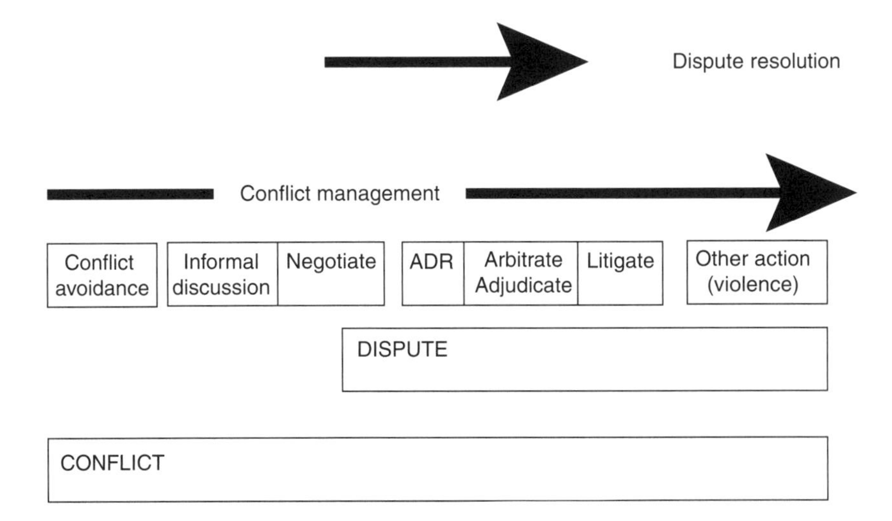

Figure 2.1 Conflict continuum

relevant where an arbitration or other dispute resolution provision in a contract provides that disputes are to be referred to arbitration or other dispute resolution.

The meaning of the word dispute would at first sight seem to be relatively straightforward. Indeed, cases such as *Hayter* v. *Nelson* (1990)[6] and *Cruden* v. *Commission for New Towns* (1995)[7] have stated that an ordinary English word such as dispute should be given its ordinary meaning. However, there is a considerable body of case law concerning the question of what constitutes a dispute. Much of that case law has been associated with arbitration and/or construction adjudication under the Housing Grants, Construction and Regeneration Act (HGCRA). Section 108 of the HGCRA provides that: 'A party to a construction contract has the right to refer a dispute arising under the contract for adjudication under a procedure complying with this section. For this purpose 'dispute' includes any difference'.

It is extremely common in construction adjudication[8] for the responding party to allege that it has not previously been given the opportunity to review the case put forward by the referring party and, therefore, that there is no dispute capable of being referred to construction adjudication. On the basis of this, the responding party will contend that the adjudicator does not have jurisdiction to deal with the matter. In 2003 at least four cases were pursued on this point alone.

The case of *Cowlin Construction Ltd* v. *CFW Architects*[9] considered the question of definition, and, in doing so, the court provided a useful summary of the relevant cases.

It appears from case law that, while there is no special meaning to be given to the meaning of the word dispute, there are certain factors to take into consideration when deciding whether or not there is a dispute. The approach adopted by the courts is one that attempts to prevent one party from ambushing the other party. There seem to be two schools of thought as to what is required for the crystallisation of a dispute. The wide approach advocated by *Halki Shipping Corporation* v. *Sopex Oils Ltd*[10] is where a claim made and not admitted is sufficient (*Cowlin*, *Costain* v. *Wescol Steel*[11] and *Orange EBB v ABB*[12]). The narrow approach advocated by *Carillion* v. *Devonport*[13] and *Beck Peppiatt* v. *Norwest Holst Construction*[14] shows a reluctance to allow ambushes and for dispute resolution to be commenced prematurely.

The myriad of techniques used to resolve disputes

There has been considerable recent interest in dispute resolution; particularly as a means of making savings by optimising efficiency in dispute resolution. In fact, the ADR Pledge (alternative dispute resolution), relaunched in 2011 as the 'dispute resolution commitment', is a central tenet of the UK government's commitment to greater efficiency in dispute resolution.

The following section reviews the main techniques available (used) and compares three key techniques. This is by no means an exhaustive or exclusive list; there are undoubtedly others, indeed one definition of ADR is 'appropriate dispute resolution', and there may be a 'killer application' yet to be devised.

The range of conflict management and dispute resolution techniques include the following:

- *Conflict management/dispute avoidance* – incorporates a variety of techniques some used consciously and some subliminal to avoid the escalation from normal conflict into dispute. Examples might include: risk management to ensure that risks are identified, analysed and managed; procurement strategies to ensure that risks are appropriately allocated; and contractual arrangements to allow sensible administration. Specific examples include: clearer project definition; equitable risk allocation; improved procurement and tendering procedures; and partnering or relationship contracting.
- *Negotiation* – easily the most common form of dispute resolution, carried out in many forms everyday by just about everybody. In negotiation, the parties themselves attempt to settle their differences using a range of techniques from concession and compromise to coerce and confront.
- *Mediation* – a private and non-binding form of dispute resolution where an independent third party (neutral) facilitates the parties, reaching their own agreement to settle a dispute. Mediation is often a structured process where the settlement becomes a legally binding contract.
- *Conciliation* – a process of mediation where the neutral proposes a solution. In the same way that we distinguished between a continuum of conflict and dispute, a continuum of mediation and conciliation shows mediation at one facilitative end and conciliation at the other evaluative end of the continuum.
- *Med-arb* – a combination of mediation and arbitration where the parties agree to mediate, but if that fails to achieve a settlement the dispute is referred to arbitration. The same person can act as mediator and arbitrator in this type of arrangement.
- *Dispute resolution adviser (DRA)* – the concept of DRA is the use of an independent intervener. This independent intervener is paid for equally by the employer and the contractor to settle disputes as they emerged, rather than wait until the end of the contract.
- *Dispute review boards (also dispute review panel, dispute avoidance panel, dispute adjudication panels)* – a process where an independent board evaluate disputes.

- *Neutral evaluation* – a private and non-binding technique where a third, neutral party (often legally qualified) gives an opinion on the likely outcome at trial as a basis for settlement discussions.
- *Expert determination (also submission to expert, reference to an expert, expert adjudication)* – long-established procedures in English law and have been used across a number of industries. Examples include: accountants valuing shares in limited companies; valuers fixing the price of goods; actuaries carrying out valuations for pension schemes; certifiers of liability for on-demand performance bonds; and Adjudicators who are said to be acting 'as expert and not as arbitrator'.
- *Mini-trial (or executive tribunal)* – a voluntary non-binding process. The parties involved present their respective cases to a panel comprising senior members of their organisation. The panel is assisted by a neutral facilitator and has decision-making authority. After hearing presentations from both sides, the panel ask clarifying questions and then the facilitator assists the senior party representatives in their attempt to negotiate a settlement.
- *Construction adjudication* – refers to Statutory Adjudication in Construction Disputes as set out in the HGCRA 1996. Here, decisions of an adjudicator are binding on the parties at least until a further process is invoked (Arbitration or Litigation).
- *Arbitration* – a formal, private and binding process where disputes are resolved by an award of independent tribunal (third party or parties, the arbitrator or arbitrators). The tribunal is either agreed by the parties or nominated by a further independent body; for example, a court or professional institution.
- *Litigation* – a formal process whereby claims are taken through court and conducted in public. Judgements are binding on the parties subject to rights of appeal.

Each of these, with the exception of litigation, are considered later.

The stages of conflict management and dispute resolution

The stages of conflict management and dispute resolution are usefully described in a document produced by the Office of Government Commerce, *Dispute Resolution Guidance*.[15] The stages are:

- *Stage 1* – negotiation.
- *Stage 2* – non-binding techniques and processes.
- *Stage 3* – binding techniques and processes.

This epitomises the current approach which seeks to make savings by optimising efficiency in dispute resolution. There is strong support for this in the UK by Government and internationally by research teams, such as those at Harvard[16] and Cornell.[17]

The principal stages and the dispute resolution options are shown in Figure 2.2 and Table 2.1.

Comparison of litigation, with construction adjudication, arbitration and mediation

It is useful to compare and contrast the major dispute resolution techniques in areas where the characteristics of each technique are highlighted. Litigation, construction adjudication, arbitration and mediation are compared under the following headings:

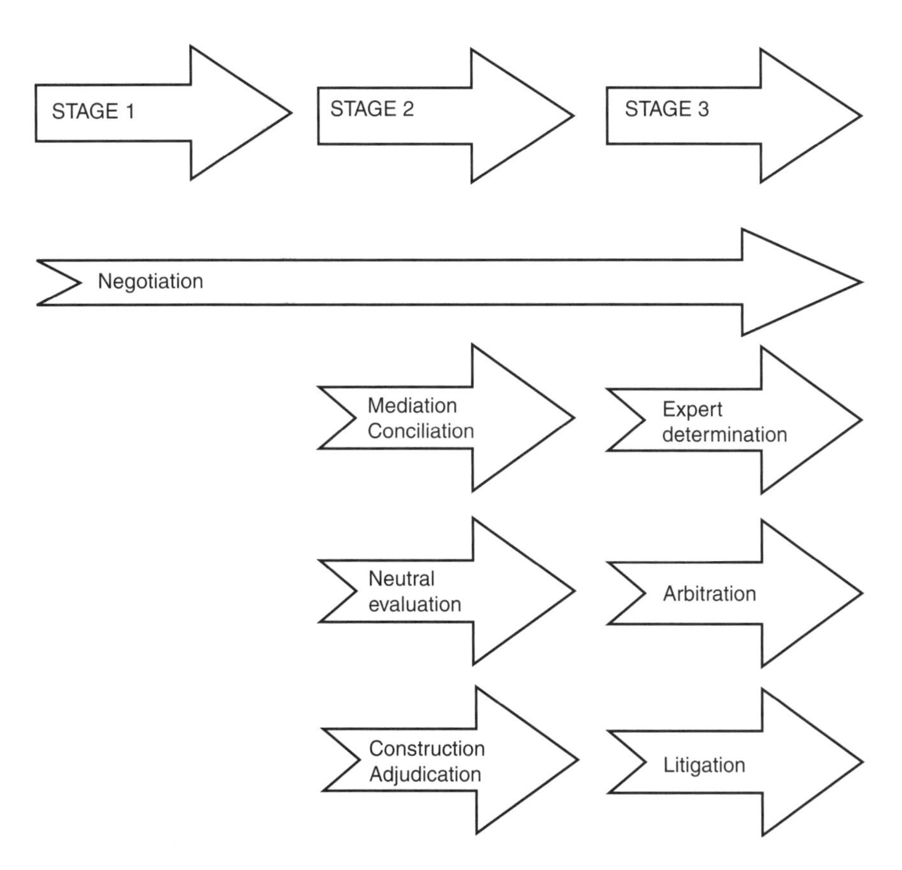

Figure 2.2 The principal stages of dispute resolution

Table 2.1 Dispute resolution options

Method	*Common law/ statute basis*	*Frequency of use*	*Speed*	*Cost*	*Confidentiality*	*Binding*	*Adversarial*	*Special features*
				Stage 1				
Negotiation	No	Very common	Varies	Low	Yes	No	No	Can continue throughout the dispute
				Stage 2				
Mediation	No	Common	Fast	Low	Yes	No (unless agreed)	No	–
Conciliation	No	Fairly common	Fast	Low	Yes	No (unless agreed)	No	Often included with mediation
Neutral evaluation	No	Infrequent	Fast	Low	Yes	No	No	–
Adjudication	Yes	Common	Fast	Low	Yes	Yes (until completion or Abr/Li)		Statutory adjudication is construction specific
				Stage 3				
Arbitration	Yes	Common	Contingent	Contingent	Yes	Yes	Yes	–
Expert determination	No	Fairly common	Fast	Moderate	Yes	Yes	Yes	–
Litigation	Yes	Common	Slow	High	No	Yes	Yes	–

- Formality.
- Speed.
- Flexibility.
- Cost.
- Confidentiality.
- Relationships.
- Control and choice.
- Solutions.

Formality

Mediation is an informal process; the parties may agree to certain mediation rules but they are at liberty to amend any rules. It is often said that the parties are in control of the settlement, but the mediator is in charge of the process. There is no requirement to produce specified information before the mediation can commence, neither is there a requirement to spend resources filing and serving documents. Mediation is informal and uncomplicated. Construction adjudication is an informal process and the procedure is, within certain bounds, at the discretion of the adjudicator. Arbitration has been criticised for mimicking litigation; many steps have been taken to redress this and arbitrations are less formal. Nevertheless, arbitration might be considered formal and complicated when compared with mediation. Litigation is, properly, a highly formalised process with specialised rules; non-compliance may prevent litigation proceeding. Resources have to be committed in filing and serving documents. Litigation is a highly formal and complicated process.

Speed

In mediation the timing is within the control of the parties. Subject to the availability of suitable and acceptable mediators, mediation may take place as quickly as the parties desire. The length of the mediation is similarly in the control of the parties; they can agree to stay as long, or as briefly, as required. The great majority of mediations are restricted to one working day or less. Construction adjudication operates under very tight timescales laid down by legislation. The maximum time from notice to decision is 35 days, which may be extended by agreement to 49 days. Speed is often claimed as a feature of Arbitration; however, the reality is that the availability of all the parties involved, not least the arbitrators, dictate that the process is often protracted. Litigation is often an infuriatingly slow process – in many jurisdictions advisors talk in terms of years rather than months, as with the timescale for trial dates. Although great strides have been taken in many countries to address this, in the UK the Civil Procedure Rules following the Woolf Review of Civil Justice is a particular example, time continues to be an issue.

Flexibility

Mediation is a flexible process and all arrangements can be changed if it becomes apparent that this is necessary. Arbitration can share much of this flexibility and the 1996 Arbitration Act has given arbitrators wide-ranging powers to achieve flexibility. Adjudicators, too, have much scope for flexibility. Litigation is an inflexible process, specific steps must be taken to initiate and progress matters.

Cost

Mediation is an inexpensive process, achieved and facilitated by the informality and speed of the process. The amount of lawyer involvement can be reduced if the parties agree and, in many cases, the cost of preparing for mediation is marginal to the other preparation. The parties can share the costs associated with the mediation in an agreed fashion. Construction adjudication can be an inexpensive process as a result of the tight timescales. Arbitration can certainly help in reducing costs and dealing with a dispute in a proportionate manner. In comparison to litigation it must be remembered that, while the state pays for the majority of the courts' and judge's costs in many cases, in arbitration the parties must pay the arbitrator's costs. Litigation is an expensive process, dictated by the formality and slowness of the process. There are many, many examples of the disproportionate costs of litigation. Among the most famous is the Dickens' *Bleak House* example of *Jarndyce* v. *Jarndyce*, where the parties disputing a will expended the entire legacy in legal costs when they disputed the terms of the will!

Confidentiality

Here things are clear: in construction adjudication, arbitration and mediation all matters are confidential. This is an important issue for commercial disputes where the parties often wish to avoid publicity and to keep commercial confidentialities. There is often an issue where arbitration awards are the subject of appeal or referral to the courts, in which case all matters will become public. Litigation is a public matter, and although civil commercial litigation seldom attracts tabloid press interest, it is clear that litigation can expose confidential issues.

Relationships

Again, a clear difference: mediation is a non-adversarial process while litigation and arbitration are both adversarial. Construction adjudication may avoid the dysfunctional aspects of adversarialism. In facilitative mediation (see Chapter 5) the parties do not seek to convince the neutral that they are in the right, or that others are in the wrong. The emphasis of facilitative mediation is on the parties' interests as opposed to parties' rights. As a result, mediation need not affect working relationships in an adverse manner. Sometimes mediation can improve relationships, as parties achieve an improved understanding of underlying interests and concerns. Litigation and arbitration, on the other hand, are not conducive to even maintaining relationships, let alone improving them. Opposing parties aim to convince the tribunal that the law and the facts support their argument to the detriment of the other side. This seldom helps relationships and often destroys them. Construction adjudication allows the power imbalance in relationships to be dealt with in that weaker subcontractors have a clear route to deal with more powerful contractors.

Control and choice

In mediation, the control of the dispute always remains with the parties and the choice is theirs. Who will be the mediator? Where will the mediation take place? When will it take place? Who will attend? Mediation is a voluntary process and the parties remain in control. This control means that the parties have to 'buy in' to the settlement and any resolution becomes their own settlement. Litigation, construction adjudication and arbitration hands over the dispute to the lawyers and the judge or arbitrator or adjudicator. The process passes control and choice in a similar fashion.

Solutions

The essential difference is that mediation allows for creative solutions to disputes, and during a mediation a wide range of issues can be addressed or uncovered. These issues can include past unresolved matters and even future intentions. The solutions to the issues can take many forms, and are not restricted to payment of money; they can be as creative as the parties to the dispute. Mediated agreements have included:

- Apologies.
- Future business arrangements.
- Revamped commercial arrangements.

Litigation and Arbitration cannot allow for creative solutions but must be limited to the legal remedies available. Construction adjudication is similarly restricted by legal remedies but does allow prompt solutions which permit the project to be completed.

The government's pledge to ADR

In the UK the result of this comparison of techniques, and a desire to make savings by optimising efficiency in dispute resolution, was that on 23 March 2001 the Government made a pledge that government departments will only go to court as a last resort. Instead, they will settle their legal disputes by mediation (or arbitration) whenever possible. Government departments and agencies will settle legal cases by ADR techniques in all suitable cases whenever the other side agrees. The pledge is worth repeating in full:[18]

> *Settlement of government disputes through alternative dispute resolution*
>
> Government departments and agencies make these commitments on the resolution of disputes involving them:
>
> - Alternative dispute resolution (ADR) will be considered and used in all suitable cases wherever the other party accepts it.
> - In future, departments will provide appropriate clauses in their standard procurement contracts on the use of ADR techniques to settle their disputes. The precise method of settlement would be tailored to the details of individual cases.
> - Central Government will produce procurement guidance on the different options available for ADR in Government disputes and how they might be best deployed in different circumstances. This will spread best practice and ensure consistency across Government.
> - Departments will improve flexibility in reaching agreement on financial compensation, including using an independent assessment of a possible settlement figure.
> - There may be cases that are not suitable for settlement through ADR; for example, cases involving intentional wrongdoing, abuse of power, public law, human rights and vexatious litigants. There will also be disputes where, for example, a legal precedent is needed to clarify the law, or where it would be contrary to the public interest to settle.
> - Government departments will put in place performance measures to monitor the effectiveness of these undertakings.
>
> 23 March 2001

The requirement to put in place monitoring measures was felt by many to be the crucial issue for government departments. Not only were they encouraged to use the pledge but their use would be monitored. The monitoring reports make interesting reading; the latest report at March 2010 states: 'During the reporting period 2008/09, ADR has been used in 314 cases with 259 leading to settlement, saving costs estimated at £90.2m'.[19]

In addition to this monitoring, the principle that litigation should be a last resort has the approval of the courts. In *Frank Cowl and other* v. *Plymouth City Council*,[20] Lord Woolf said: 'insufficient attention is paid to the paramount importance of avoiding litigation whenever this is possible'.

Why commercial projects go wrong: the construction example

It certainly seems that commercial projects do go wrong, everyone knows that, it is one of the problems of commercial. As an example we might consider the construction industry in the UK where the problems are writ large. These problems have intrigued, one might say obsessed, the industry and government for 50 years. Reports on construction are nothing new and even an incomplete list of reports since the Second World War makes depressing reading.

A simplistic analysis of all these reports finds three key areas. Construction has a reputation for producing products which are not to the quality expected, over budget and over programme. What is the problem with construction? That much is easy, the industry has a poor image and is renowned for products which are:

- Poor quality.
- Over budget (expensive).
- Over programme (late).

Stella Rimington tells this with stunning clarity in her book, *An Open Secret*[21], about her time in MI5: 'Like all huge building projects, particularly in the public sector, the Thames House Refurbishment had been fraught with difficulties ... It was clear that dealing with the building industry was just as tricky as dealing with the KGB'.

Construction projects are poor quality, late and expensive. Another feature is the preponderance of disputes, often described as 'adversarial attitudes'. Many would question the evidence for any of these problems, but here let's restrict the questioning to the disputes issue.

The adversarial attitudes problem is often expressed as received wisdoms about construction. Simply put, there exist received wisdoms that:

1 Construction suffers more contractual disputes than other industries.
2 The occurrence of disputes has risen recently and continues to rise.
3 The performance of the industry is adversely affected by the disputes.

The received wisdoms are repeated throughout the construction literature. For a contemporary confirmation in the UK see Latham and National Audit Office;[22] the same is repeated internationally.[23] However, there is little empirical work to test the received wisdoms, and random theorising is allowed to pass unchallenged. It is intriguing that in an area accustomed to the rigours of evidence, both legally and scientifically – the law and academia, this has been allowed to pass with barely a murmur of protest. This evidential sloppiness peaked during the debate which accompanied the introduction of Housing, Grants Construction and Regeneration Act in the UK, and the introduction of construction adjudication, when the Department of the Environment (1996) claimed *inter alia* that 'there is compelling anecdotal evidence that adjudication would reduce overall project costs'.[24] It is suggested that the phrase 'compelling anecdote' is an oxymoron.

Disputes and conflict on projects, or contracts, are unpleasant. They divert valuable resources from the overall aim, which must be completed on time, on budget and to the quality specified. In addition, they generally cost money, take time and can destroy relationships which may have taken years to develop. Commerce and its management needs to recognise both conflict management and dispute resolution.

Explanation or prediction of commercial disputes

That lots of the literature states that dispute avoidance is to be preferred is self-evident. One clear example: 'The best solution is to avoid disputes'.[25]

If we deal with this issue of avoiding disputes, if we seek to avoid disputes, it is axiomatic that we seek to predict, because by prediction we can take the necessary evasive action. Prediction is at the very heart of the scientific method of research and we might look to other disciplines for guidance on predictive research. Medicine is an example a mature scientific discipline where predictive techniques have developed a concept of preventative medicine. Medical science has a well-established branch of aetiology: the study of the causes, for example, of a disease. The word aetiology comes from the Greek *aitia*, a cause, plus *logos*, a discourse.

An aetiological approach might throw new light on construction disputes. Continuing the analogy with medicine, construction disputes represent the dysfunctionality of conflict and, therefore, the disease on the body of construction. Aetiology has shed new light on many problems of medical disease, providing important clues to the understanding of the nature of the disorder and promoting advances in diagnosis, treatment and prevention.

Diagnosis is the act of identifying a disease from its symptoms or signs. In construction disputes, the symptoms of dispute are not currently considered, only when a dispute has manifested itself are the parties concerned with the dispute and then only with the resolution of that dispute. Treatment of the dispute, following most dictionary definitions and not the medical concept, is the application of the techniques or actions in specified situations. Treatment of disputes is, therefore, the application of the techniques of conflict management and dispute resolution. Prevention is the act that is used to avoid disease and, in the argument presented here, prevention is made possible by the prediction of the occurrence of disputes.

So the suggestion is for a predictive model. Why do disputes arise? And when? And what causes them? If this could be forecast then disputes could be avoided and, if not, the most efficient technique could be used for their resolution. This fits in with the existing theory that conflict is all around and dysfunctional conflict becomes dispute.

A traditional student essay might be: 'Conflict and dispute are inevitable. Discuss.' The Marxist view is that conflict is an inevitable part of the struggle for scarce resources. The Popperian[26] view is that we cannot say whether or not disputes are inevitable since this requires a prediction of the future, and we cannot predict the future. Or, as Neil Bohr is alleged to have said, prediction is difficult particularly if it involves the future.

The cost of conflict and dispute

In the wider academic field there is considerable interest in the cost of conflict. Nobel Prize winner James Stiglitz is one of many who have considered the cost of conflict a tool which attempts to calculate the price of conflict to the human race.[27] But what of the cost of conflict and disputes to industry and commerce? In November 2008, Lord Justice Jackson was appointed to lead a fundamental review of the rules and principles governing the costs of civil litigation and to make recommendations in order to promote access to justice at proportionate cost. His findings were published in January 2010.[28] What of costs as a heuristic? Is there a simple estimate (perhaps by a percentage) of costs of, say, litigation? Such analyses are difficult to find, but they have been attempted.

Lord Woolf in his report[29] includes a study of costs in Official Referees' Cases, while Fenn and Black (1999)[30] indicate that costs may lie in the range of 19–25 per cent of the amount in dispute (where this amount exceeds £100,000). Fenn and Black also indicate that perhaps six per cent of construction contracts produce disputes of such severity that the courts or arbitrators are involved. As a guide, therefore, if the construction industry output in 2009 is taken at £123bn[31], then perhaps £7.4bn is in dispute which generates costs in the range of £1.402–1.845bn. The potential for improvements in construction is obvious, acknowledging the simplicity of the argument and calculation. £1.402–1.845bn which might be spent on construction disputes either by clients or contractors adds to the price of construction works. The £1.402–1.845bn is either added to clients' costs or to contractors' prices. In the context of the recent Comprehensive Spending Review and the current economic climate, it's more pertinent than ever to be looking at this.

Whatever the cost of dispute, the cost is something that should be avoided.

The intriguing case of the dabbawalla

The Hindi word dabbawalla literally means 'person with a box'. The term is also used to describe a person in the Indian city of Mumbai whose job it is to collect and deliver freshly cooked lunches to office workers, often in the suburbs. Sometimes the term 'tiffin wallah' is used in place of dabbawalla, as tiffin is an obsolete word (in English but not in Indian English) for a light meal.

Each day in Mumbai up to 200,000 lunch boxes are moved, from home to office and the empties back again by up to 5,000 dabbawallas. The fee paid to the dabbawalla is nominal, and most dabbawallas are illiterate. Mistakes, and therefore conflict or dispute, are almost unheard of and it is estimated that less than one in six million deliveries produce a mistake.

Mumbai is one of the most densely populated cities in the world and long journeys to work by its inhabitants are commonplace. Many Indian office workers prefer to eat home-cooked food at lunchtime, but it would be impossible for them to travel home during their break. A system developed whereby lunch boxes from home containing food are collected in the morning, delivered to the office workers at lunchtime and the (same) empty lunch boxes returned to home in the afternoon.

The system works like this:

1 A collecting dabbawalla collects dabbas (boxes) from home, which are then taken to a sorting place where they are bundled and put on trains for the city. Dabbas are marked with a colour or symbol.

2 The train dabbawalla transports the dabbas to the city and delivers them to a local dabbawalla.

3 The local dabbawalla delivers the dabbas to the offices.

In the afternoon, by the reverse process, dabbas are delivered back to the homes in the suburbs for use the next day. Dabbas are mostly stackable circular metal containers. Each dabba has a unique mark indicating its origin, travel route and destination. Each box can change hands many times and travel on many trains in the course of its daily journey.

If you subscribe to the view that conflict and dispute are inevitable, how then do you explain the dabbawalla? By any standard a dispute ratio of one in six million is effectively no disputes – it is in the noise. There has been considerable interest in the dabbawallas from academics and industry. Harvard business review[32] pointed to the illiteracy rate of the dabbawallas and wondered if high-tech approaches might learn from the simplicity and elegance of the colour coding and symbol approach that the illerate dabbawallas had produced. Forbes magazine found the dabbawalla's reliability to be that of a six-sigma standard. A six-sigma process is one in which 99.99966 per cent of the products manufactured are statistically expected to be free of defects.

Some have pointed to the organisational structure of the dabbawallas' collective as an explanation of the success and the absence of disputes. Everyone who works within this system is treated as an equal and working to the same values. Regardless of a dabbawalla's function, everyone gets paid the same and works to towards the same outcome. If we return to Marx's conflict theory, the basic reason for dispute is removed.

3 Conflict management and dispute avoidance

Pink ribbon

This chapter uses the terminology, or taxonomy, developed in Chapter 2, and talks of conflict management and dispute avoidance, rather than conflict avoidance.

Conflict management and dispute avoidance are used in a variety of techniques, some used consciously and some subliminally to avoid the escalation from normal conflict into dispute. Examples include: risk management to ensure that risks are identified, analysed and managed; procurement strategies to ensure that risks are appropriately allocated; and contractual arrangements to allow sensible administration. Specific examples include: clearer project definition; equitable risk allocation; improved procurement and tendering procedures; and partnering and relationship contracting.

It might be argued that this chapter is far and away the most important chapter in the book, and that conflict management and dispute avoidance are more important than dispute resolution. It is clear just by looking at the range of books on dispute resolution that general interest is in dispute resolution. Perhaps this can be explained by the fact that conflict management and dispute avoidance are obvious things that don't need to be stated. Conflict management and dispute avoidance is project management, is human nature, is done all the time. Well, if that is the case, it could be done better.

A case study is made of Terminal 5 at London Heathrow Airport where acceptance of risk by the client resulted in dispute avoidance.

Introduction (procurement and contracts)

Some would say that procurement and contracts are central to conflict management and dispute avoidance. But before starting on the bewildering language of procurement it is worth considering contracts, because at the heart of procurement is contract.

Procurement is about obtaining goods and services required from external organisations. It breaks down into two main activities:

- Strategic decisions in terms of make or buy (see transactional cost economics); work breakdown structure and which parts of the project are allocated to what type of organisations; allocation of principle risks; how these organisations are paid and incentivised to perform; the degree of cooperation required between participating organisations and how they are selected, etc.
- The selection procedure itself to select the 'best fit' organisations for their part of the project.

If the decision is made to *buy* rather than to *make*, then a product or service will have to be purchased, and that product or service requires a contract.

Contracts

Simply contracts might be categorised as:

- Lump sum.
- Measure and value (ad-measurement).
- Cost reimbursement (cost plus).

All standard form contracts can be categorised in this way. Examples of standard forms and their categorisation include:

- Lump sum – Institution of Chemical Engineers, the red book, fourth edition, 2001.
- Measure and value – Institution of Civil Engineers, seventh edition, 2007.
- Cost reimbursable – Institution of Chemical Engineers, the green book, third edition, 2002.

Figure 3.1 tries to show these types by risk apportionment. This is greatly simplified.

- Lump sum – risk mostly with contractor.
- Measure and value – risk mostly shared client/contractor.
- Cost reimbursement (cost plus) – risk mostly with client.

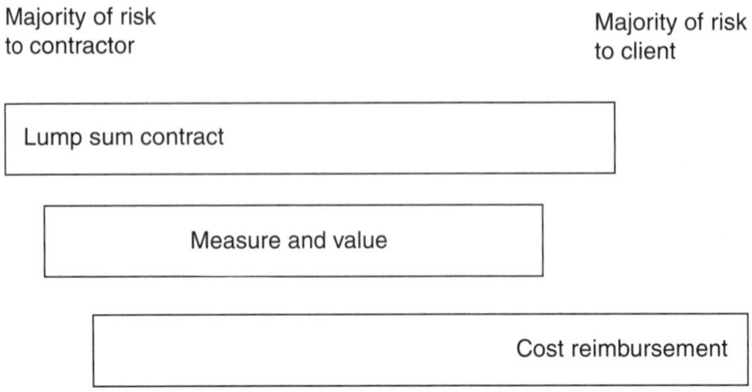

Figure 3.1 Types of contract by risk appointment

Lump sum

Here, the work is clearly identifiable and quantifiable and the extent, boundaries and detail of the work are known.

- Example:
  Providing a product, goods or service where the work is clearly identifiable – extension to a house.

Measure and value

Here, although the work is clearly identifiable, the extent, boundaries and detail of the work are not clear and the quantities may change.

- Example:
  Providing a product, goods or service where the work is clearly identifiable but the extent, boundaries and detail of the work are not clear and the quantities may change – new railway line between Manchester and London.

Cost reimbursement (cost plus)

Here the work is NOT clearly identifiable or quantifiable, and the extent, boundaries and detail of the work are not clear.

- Example:
  Providing a product, goods or service where the work is NOT clearly identifiable or quantifiable, and the extent, boundaries and detail of the work are not clear – emergency work or asbestos removal from government offices.

Procurement

Procurement is the process by which the resources (goods and services) required by a project are acquired. It includes:

- The development of the procurement strategy.
- Preparation of contracts.
- Selection and acquisition of suppliers.
- Management of the contracts.

Contracts and procurement as conflict management and dispute avoidance

Good contracts and good procurement are fundamentals. Without these in place, parties may have the wrong partner, be unsure of their rights and obligations, and motivated to perform to different objectives. Consequently, the chances of a successful project are diminished and the prospects of dispute increase.

Good contracts and procurement can help drive a project towards success through selection of the right parties, under the right contract strategy which aligns the parties' motivations under clear contract terms. Therefore, contracts and procurement are central to conflict management and dispute avoidance.

Procurement, at its simplest, is how to go about obtaining the goods and services needed from external organisations. It breaks down into two main activities:

1. Strategic decisions in terms of make or buy; which parts of the project are allocated to what type of organisations; allocation of principle risks; how these organisations are paid and incentivised to perform; the degree of cooperation required between participating organisations and how they are selected, etc.
2. The selection procedure itself. This selects the organisations for particular parts of the project. This is often described as 'best fit'.

Having chosen the 'best fit' organisation, a contract then needs to put in place which reflects these strategic decisions and, above all, accurately and sufficiently describes with clarity what that organisation is wanted to deliver, by when and for how much. It also needs to describe how change, which inevitably comes to varying degrees with a project, will be assessed and implemented. This contract then needs to be managed.

In an era of technology and complexity, few organisations deliver projects, let alone major ones, with no external help. Where the goods or services are predefined or simple, the contract and procurement process is relatively easy. When procuring complex projects, with evolving needs and which may involve tangible and intangible deliverables, it becomes much more complicated.

Conflict management and dispute avoidance the techniques

There are many techniques and perhaps they are merely codification of what everyone knows: action is required to avoid conflict from escalating into dispute. Some of the techniques are used consciously and some are subliminal. Which should be chosen here? Of course it's arbitrary, and perhaps the best statement is: *It might be argued that this chapter is far and away the most important chapter in the book; and that Conflict Management and Dispute Avoidance are more important than Dispute Resolution.* Examples examined here include: risk management to ensure that risks are identified, analysed and managed; procurement strategies to ensure that risks are appropriately allocated; and contractual arrangements to allow sensible administration. Specific examples include: clearer project definition; equitable risk allocation; improved procurement and tendering procedures; partnering and relationship contracting.

Back to the argument that this chapter is far and away the most important chapter in the book, and that conflict management and dispute avoidance are more important than dispute resolution. However, it is clear just by looking at the range of books on dispute resolution that the general interest is in dispute resolution. How can this be explained? Perhaps by the fact that conflict management and dispute avoidance are obvious things and they don't need to be stated. Conflict management and dispute avoidance is project management, is human nature, is what is done all the time. If that is the case, examples around us and the disputes we see indicate it could be done better. Universities should run courses in conflict management and dispute avoidance; authors should write books on conflict management and dispute avoidance and practitioners should develop more techniques of conflict management and dispute avoidance. That the courses, books and techniques are few is an intriguing question.

Specific examples

The examples considered are: clearer project definition; equitable risk allocation; improved procurement and tendering procedures; partnering and relationship contracting.

For each example a definition is provided, the key indicator stated, the costs associated are outlined and some indication of the international status is given.

Clearer project definition

Definition: defining a project is a process of selection and reduction of the ideas and perspectives of those involved into a set of clearly defined objectives, key success criteria and evaluated risks. Key indicator: clear and accurate definition of a project is one of the most important actions available to ensure project success. Cost: there are costs associated with clearer project definition, but the benefits and savings far outweigh these costs. International Status: clearer project definition is a widely used technique in the UK, USA, Australia, Canada and other common law and civil jurisdictions. The UK Construction Best Practice Programme makes clearer project definition and better briefing one of its action areas and special interest group.

In all commercial projects the client must brief the contractor about what is expected. Often there will be a formal or written brief or series of briefs that may form part of a tender document. Briefing is the process through which the client and the contractor explore, develop and communicate the client's requirements.

Briefing should include:

- Establishing objectives and/or business case.
- Examining other means of achieving them before deciding to build or produce.
- Spending time at the beginning to define what is wanted, when and for how long, changes later are expensive.
- Establish any budget and/or time limitations.
- Prioritise time, cost and quality.
- Take care to choose the people to represent, advise and work for the project. They should be qualified, experienced and able to work well with each other.
- Identify the risks involved, quantify them and confirm budget. Identify the cost of the project over the period of intended use (whole or life cycle costing).
- Identify the options.
- Monitor progress and performance and be ready to deal with the unexpected.

Briefing takes place throughout the commercial process from project inception to completion. It is important that the client is actively involved at all stages to ensure that the project meets requirements. Critical decisions are often taken during the early stages of the project and full client participation in these is essential. Adequate time and resources must be applied in a productive way for briefing to be effective. Much depends on interpersonal and managerial skills and these must be developed to meet the demands of a particular project and set of participants. Factors such as client experience, complexity of organisation, organisational culture, rate of organisational change, project complexity and degree of project development all need to be taken into consideration. Several areas in which there is scope for improvement in briefing practice have been identified. These include optimising client's often extensive in-house commercial expertise and control over its projects, managing the project dynamics, appropriate user involvement, appropriate team building and appropriate visualisation techniques. Where a client has no expertise then appropriate expertise must be procured from outside.

Briefing may be more successful if it is approached with:

- Carefully thought out requirements.
- The essential information provided at each stage of the project.
- A flexible approach balancing the requirement for quality with the concern to control costs and meet deadlines.
- Trusting relationships.

Clear and accurate definition of a project is one of the most important actions that can be taken to ensure a project's success. The clearer the objective the more likely that it can be achieved. Defining a project is a process of selection and reduction of the ideas and perspectives of those involved into a set of clearly defined objectives, key success criteria and evaluated risks.

This definition process should culminate in the production of a project definition document, sometimes called a project charter. The project definition document should be approved and issued by a manager with the authority to apply organisational resources to the project activities. Therefore, the seniority of the manager or the management team will be commensurate with the size, cost and business value of the project. As a minimum, the project definition should include a statement of the purpose that the project seeks to address and the description of the product, service or deliverable objectives that will be its output.

One way to define a project is for the project leader to ask a standard set of questions of the project team, colleagues with particular expertise and senior managers. The questions might fall into:

1 The purpose (or mission)

   Examples:
   What is the reason for doing the project? What is the project about in broad terms? Who wants it done and why? What is its title?

2 Deliverables

The fundamental objective of a project is to deliver something new. What is being delivered?

Better briefing has been a recurring theme for the construction industry. Examples of better briefing in can be found in architectural literature and at Constructing Excellence – the single organisation charged with driving the change agenda in construction in the UK.

Equitable risk allocation

Definition: equitable risk allocation is a process where the risk is allocated to the party best able to control and manage that risk. Key indicator: equitable risk allocation has been identified as one of the strategies that would reduce the incidences of claims and disputes. Cost: the costs of implementing equitable risk allocation (e.g. the risk identification and allocation process) are recognised, but the benefits and savings far outweigh these costs. International status: equitable risk allocation is a widely used technique in many countries. It is favoured in the UK, USA, Australia, Canada and other common law and civil jurisdictions. There are few countries untouched by equitable risk allocation; although in many countries and in many sectors there remains much ignorance of the benefits to be obtained from this practice.

Once again, construction is an innovative trail blazer in this area of conflict management and dispute avoidance. Construction project management involves the planning, organising, directing and controlling of company resources for the completion of a project development. Project success is usually measured by the achievement of the time, cost, quality and maximising resource utilisation. The achievement of these objectives can also be measured by the incidences of claims and disputes and their resolution. Dispute prevention is then one of the major tasks in construction project management. Equitable risk allocation has been identified as one of the strategies that would reduce the incidences of claims and disputes.

Every risk has an associated price visible or hidden. Visible costs appear in project bids as contingency or insurance costs and can be compared. Onerous contract conditions promote hidden costs. Hidden costs (in terms of time and money) include:

- The cost of restricted bid competition.
- The cost of increased claims/disputes.
- The cost of replacing a lower quality contractor who is more likely to unknowingly accept a grossly inequitable risk allocation.

- The cost of operating an adversarial owner–contractor relationship in terms of final product quality, cooperative implementing of change order processing, reputation.
- Ultimate project outcome.

The client has an essential role in improving working relationships, contract execution and overall project performance, by the decisions made regarding risk allocation. The general conclusion is that the use of onerous contract provisions that cause the contractor to assume inequitable, unbearable and uncontrollable risks will directly and negatively impact the owner–contractor working relationship.

Beyond equitable allocation of risk, there are additional steps a client can take to improve working relationships. The development of project problem-solving teams with client and contractor's personnel to anticipate potential project problems and provide workable solutions in advance. Another suggestion is to give increased authority to the client's on-site project manager so decisions can be made at levels closer to the work.

Risk and uncertainty are inherent in projects, and the potential for damage can be limited through proactive and systematic risk management. The generally accepted international procedure is that it is necessary to allocate risks in an equitable manner between the parties to the contract to ensure successful project delivery. The problem, of course, is what constitutes equitable risk allocation? The following are common considerations:

- Which party can best control the events that may lead to the risk occurring?
- Which party can best manage the risk if it occurs?
- Is it preferable for the employer to retain an involvement in the management of the risk?
- Which party should carry the risk if it cannot be controlled?
- Whether the premium charged by the party accepting a risk is likely to be reasonable and acceptable.
- Whether the party accepting a risk is likely to be able to sustain the consequences if the risk occurs.
- Whether, if the risk is transferred, it leads to the possibility of risks of different nature being transferred back.

If these considerations are applied, it should be possible to achieve clear and realistic terms that are acceptable to the employer and on which contractors are prepared to tender at prices which do not contain contingencies for unclear terms or for significant risks which are not possible to estimate with some certainty or which are unlikely to materialise. An acceptable 'formula' for risk allocation, might run as follows. A party should bear a risk where:

- It is in her control, i.e. if it comes about it will be due to wilful misconduct or lack of reasonable efficiency or care.
- She can transfer the risk by insurance and allow for the premium in settling her charges to the other party, and it is most economically beneficial and practicable for the risk to be dealt with in that way.
- The economic benefit of running the risk accrues to her.
- To place the risk on her is in the interests of efficiency (which includes planning, incentive, innovation) and the long-term health of the industry on which that depends.
- The loss falls on her in the first instance, if the risk occurs.

The job of trying to balance the principles in practice is a hard one; but with a set of declared principles rather than undeclared, there is a standard to refer to.

Improved procurement and tendering procedures

Definition: this is a statement of intent rather than something capable of definition. Improvements to procurement and tendering will allow all the parties to manage conflict and avoid dispute whilst increasing profits or increasing returns to society. Key indicator: the key indicator is improved procurement and tendering procedures; this does not necessarily mean that lowest initial tender price equates with best value. Cost: there are many direct costs of improved procurement and tendering procedures (e.g. the administration of the procedures), but the benefits and savings far outweigh these costs. International status: improved procurement and tendering procedures are the aim and objectives of many countries. There are schemes in the UK, USA, Australia, Canada and other common law and civil jurisdictions.

Throughout the world, industries demonstrate similar structures, e.g. in developed economies construction accounts for 10–15 per cent of GDP and in developing economies often more than 30 per cent. In either case the industries are often fragmented, i.e. design and production are separated and many small firms or organisations exist to service a few large enterprises. The proportion of public sector expenditure varies by country but remains high, even in economies which have sought to reduce public expenditure. As a measure of subcontracting, the countries within the European Union are dominated by small and medium enterprises such that the entire construction industries of each nation are composed of small firms (less than ten employees) and the majority of the work is carried out by subcontractors. The situation is the same in many other countries. In many countries it is not uncommon for large contractors to employ no site staff at all and to subcontract entire operations.

Public sector procurement in many nations remains dominated by competitive tender where the lowest bid wins. There are many developments to avoid such systems because of their inefficiencies, but the requirements of public accountability and probability make change slow. In addition, the structural problems of industries dominated by small companies produce conservatism.

Finance: many nations have attempted to reduce public expenditure and therefore reduce taxation. In order to maintain public services these countries have a need for private finance into public schemes. A variety of procurement systems have developed to assist this: design, build, fund and operate (DBFO); build operate own (BOO); build operate own and transfer (BOOT); public finance initiatives (PFI); and public–private partnerships (PPP). The structural problems with industries described above hinder the efficient use of such schemes and steps have to be taken to remove the structural inefficiencies. In the UK construction industry, for example, these steps are coordinated by the Construction Excellence programme and include supply chain solutions. Many of these developments are mirrored elsewhere.

The move away from lowest price tendering is seen by many as a major development and in many countries there have been developments which allow the assessment of tenders on criteria other than just on price.

Partnering and relationship contracting

Definition: partnering is very simple in concept. It is just people working together – a voluntary system of handling normal, everyday problems in a mutually agreeable manner before they turn into major issues that create disputes. Partnering can be either strategic (long term) or project (short term). Some consider that strategic partnering is inappropriate for public sector or government contracts because of accountability issues, the need to share work among government service providers and secure low pricing through competitive tendering. Key indicator: partnering is clear conflict management and dispute avoidance. The partners share a common goal to achieve project success. Effective partnering requires the use of skilled facilitators to break down existing barriers and preconceptions. Cost: there are additional costs associated with partnering (e.g. a facilitator is often used), but the benefits and savings where partnering is successful far outweigh these costs. International status: partnering is a widely used technique in many countries. It is favoured in many states of the USA, Canada, and Australia and in the UK. There are few countries untouched by partnering, although in many countries and in many sectors there is limited knowledge or experience of how partnering works.

It is usual to make a split between strategic partnering (long-term partnering) and project partnering (short-term partnering). Strategic partnering can raise many problems for public sector works. Partnering can be defined as an informal process bringing together all the parties in a collaborative effort. They meet on a regular basis to review progress and deal with any problems and potential disputes from the moment they become apparent. The focus is on conflict management and dispute avoidance. The building of the partnering team may be facilitated by an outside facilitator. The US Army Corps endorsed partnering in 1990, saying:

> Clearly, the best dispute resolution is dispute prevention ... By taking the time at the start of the project to identify common goals, common interests, lines of communication, and a commitment to cooperative problem solving, we encourage the will to resolve disputes and achieve project goals.
>
> (Army Corps of Engineers, Policy Memorandum 11, 7 August 1990)

It is important to note that partnering is a voluntary system. Some ask why partnering cannot be made a requirement, but this is against the entire spirit of partnering – if it is not a voluntary agreement by all parties, it is just another contract provision. This does not downplay the importance of the contract through partnering, rather it attempts to ensure that all players with valuable experience to share are allowed to participate cooperatively in the process.

Partnering attempts to establish a working relationship among all team members based on cooperation and teamwork and achievement of mutual goals and objectives. Partnering is a concept that every contract has an implied covenant of good faith and fair dealing, and through the exercise of that agreement, the stakeholders strive to create a synergy of purpose to solve problems for the good of the project.

The project partnering process creates a new team-building environment, which fosters better communication and problem solving, and a mutual trust between the participants. These key elements create a climate in which issues can be raised, openly discussed, and jointly settled, without getting into an adversarial relationship. Through this process of teamwork and problem solving on a project, key goals are set. The partners want the quality of the work to be right the first time, the project to be completed on time, the final cost to be within budget, and disputes/litigation to be minimised.

The real impetus for partnering is the fact that the people involved in implementation have discovered that it works. A review of cases in which partnering has been used shows dramatic time and cost savings.

Relationship contracting

Definition: relationship contracting is a process to establish and manage the relationships between parties that aims to remove barriers, encourage maximum contribution and allow the parties to achieve success. Key indicator: relationship contracting is based on achieving successful project outcomes. There are some core values or guiding principles: commitment, trust, respect, innovation, fairness and enthusiasm. Cost: there are additional costs associated with relationship contracting (e.g. the facilitator), but the benefits and savings far outweigh these costs. International status: relationship contracting is a widely used technique in some countries, notably in Australia. Serious concerns remain about the use of relationship contracting in public sector projects on grounds of public accountability.

Relationship contracting is based on achieving successful project outcomes, including:

- Completion within cost.
- Completion on time.
- Strong people relationships between the parties resulting from mutual trust and cooperation, open and honest communication.
- Optimum project life cycle cost.
- Achieving optimum standards during construction and project lifetime for:
    - safety
    - quality
    - industrial relations
    - environment
    - community relations.

Relationship contracting establishes a working relationship, which is designed to deliver optimum commercial benefits to all the parties. It is founded on the principle that there is a mutual benefit to the client and the contractor to deliver the project at the lowest cost. To achieve this, the relationship between the client and the contractor cannot be taken for granted. Even if they have worked together before and have established a close working relationship it is still crucial that they build the relationship for each specific project. In order to do this the relationship must be founded on strong mutually held core values and guiding principles. These are summarised as core values and guiding principles.

- Commitment: total commitment to achieving the project goals which is actively promoted by senior management of all parties.
- Trust: to work together in a spirit of good faith, openness cooperation and not to seek to apportion blame.

- Respect: the interests of the project take priority over the interests of the parties.
- Innovation: to couple innovative or breakthrough thinking with intelligent risk taking to achieve exceptionally good project outcomes.
- Fairness: to integrate staff from all parties on the basis of fairness and the best qualified for the job.
- Enthusiasm: to engender enthusiasm for professional duties and the project's social activities.

An alternative view of conflict management and dispute avoidance

Remember that definitions are crucial but there are other views on conflict management. These mostly involve implementing strategies to limit the negative aspects of conflict. Again, most conflict management seeks to enhance learning and group outcomes management effectiveness or performance in organisational setting. Conflict management is not concerned with eliminating all conflict or avoiding conflict. Conflict can be valuable to groups and organisations.

If you like, conflict can be good and therefore needs managing; dispute is always bad and must be avoided. The classification of conflict offered in Chapter 2 is of functional and dysfunctional conflict. Dispute might be considered dysfunctional conflict.

Dispute avoidance, avoiding disputes, conflict management: the Terminal 5 case study[1]

London Heathrow Airport is the largest airport in the world. It is a major part of the UK economy: 155,000 people work there, or earn their living from it, and 68 million passenger pass through it each year. A new terminal was required, and two decades of planning, design and construction resulted in the opening of the Terminal 5 project on time, on budget and safely. What was behind it? At the time, the construction industry had a pitiful reputation for success on major projects. On the one hand, groundbreaking management thinking and the lessons learnt from leaders, client and integrated supply chain teams, which involved over 50,000 people from 20,000 companies. On the other hand, nothing more than straightforward dispute avoidance. A different commercial contract and approach by the client, BAA, enabled the construction phase to go to plan, and to be opened a year early in 2009, saving one billion pounds.

Many adjectives were used to describe BAA's Terminal 5 (T5) programme at Heathrow Airport: 'a mega project', 'enormous', 'epic', 'historic', 'huge' and 'massive'. At the time of its construction it was Europe's largest and most complex construction project, costing £4.3bn.

Many clients, when faced with a project as complex and challenging as T5, would have adopted the well-established delivery approach common in the construction industry. However, driven by a desire to reduce the costs of providing its airport facilities, BAA concluded that it could improve T5's delivery by adopting emergent project, risk and contract management methodologies.

BAA chose to manage the project itself and to accept risk rather than contract a company to manage it for them, which is the common established approach.

To execute this, retaining and managing the risk in addition to adopting a cost reimbursable form of contract, BAA required a large, highly proficient internal project management team (some have described this as acting as an intelligent client). Members of the project management team (in excess of 150 people!) took an active role in the management of each integrated (delivery) team. This concept had its roots in earlier BAA procurement, in the 1990s when BAA developed partnering agreements, using framework agreements, which incorporated integrated team working. BAA entered into many construction and consultancy agreements by the mid-1990s. The model agreement had resulted in enhanced project predictability and repeatability.

In the 1990s, BAA carried out research on large infrastructure projects in UK construction. They looked at the out-turn performance for things such as the Channel Tunnel, Jubilee Line extension, the British Library, Scottish Parliament and West Coast Main Line. In addition, between 2000 and 2002, BAA analysed every UK construction project (exceeding £1bn in value) constructed during the preceding ten years, plus every international airport project completed in the previous 15 years. As part of this study BAA investigated project processes and organisation, and the influence of individual behaviour on project performance. The research established the following:

1 Not one UK construction project (within the set parameters) had been delivered on time, on budget, safely or met its specified quality standards. Based on their analysis BAA predicted:

   > Terminal 5, a five-year build programme would probably be about two years late, a cost target of £4.3 billion would probably be at least £1–1.5 billion over budget, the quality would be variable and, statistically, 12 people would die on site. None of those consequences were acceptable to BAA.
   >
   > (Matthew Riley, Supply Chain Director, BAA)[2]

2 None of the international airport projects studied had functioned as designed when initially opened.

3 All the projects studied had experienced significant contractual and financial difficulties, and on each one the client had incurred immense reputational damage.

The study concluded that failure on projects the size of T5 was due to two main reasons:

1. *Cultural confusion* – organisational and management issues arising from ill-defined project parameters and the failure of purchasers to appreciate the needs of supply chains.
2. *The reluctance to acknowledge risk* – rather than identifying and apportioning risk appropriately at an early stage, traditional contracts generally sought to transfer risk to the supply chain, which often resulted in lengthy legal disputes when supplier performance did not meet the client's aspirations. Forty per cent of the cost of claims are the legal expenses.

(Matthew Riley, Commercial Director T5, BAA)[3]

Alternatively, project success, came from:

- Project culture.
- Effective leadership.
- Supplier 'behaviour'.

BAA thought that supplier behaviour, both positive and negative, was predominantly influenced by conditions of contract and anticipated profit margins. And BAA concluded that a step-change in construction procurement best practice was essential if T5 was to be successfully delivered. The strategy for the step change was base on four key principles:

1. *The client always bears and pays for the risk*

   Irrespective of the contractual arrangement adopted it is impossible to transfer risk:

   > We realised that, to expose waste and manage the performance more efficiently, we would have to actively hold all the risk. We realised you cannot transfer corporate risks around that are so intrinsic to the success of your company; risks that relate to the city or to airlines or regulators or to your corporate citizenship. Those risks can't be transferred down a contract. You're kidding yourself if you think they can, because, in each of those examples we looked at, there were very few suppliers that went out of business as a consequence of those project failures. The risk ultimately comes back to the client organisation.
   >
   > (Matthew Riley, supply chain director, BAA)

2 *BAA retained full liability for all project risks*

> We had to have a strategy that was, at its highest level, BAA holding all the risk all the time, and in return we expected our suppliers to come together as partners and work in an integrated team or teams. They came together to deliver projects or products, and the financial consequences of risks were underpinned by insurance policies that BAA took out directly with the market, on a strictly no-fault basis.
> (Matthew Riley, supply chain director, BAA)

3 *Suppliers' profit levels were predetermined and fixed*

BAA ring-fenced its suppliers' profits and incorporated a gain–share arrangement, whereby efficiency savings could be translated into higher margins.

4 *Partners are worth more than suppliers*

BAA implemented an 'integrated project team approach' embraced in a 'delivery team handbook', which sought to create an appropriate environment for team working with the objective of motivating, organising and generally getting the best out of the talented people working on the project.

The T5 contract

The main objective of the contract employed was to create a unique relationship under which BAA retained all the risk relating to the project. Additionally, the contract needed to be flexible as BAA appreciated that their requirements would change during the course of the contract. The contract was designed to enable all participants to concentrate on:

- The root cause of problems, not their effects.
- Working within integrated teams to deliver success in an uncertain environment.
- The proactive management of risk rather than the avoidance of litigation.

Disputes

As a result of BAA selecting a cost reimbursable contract, which incorporated pre-emptive risk management, integrated teams and promoted a non-adversarial approach and no-blame culture, there have been no reported payment disputes on the project.

During the first week T5 was open, many flights were cancelled and many more items of luggage mislaid. Why the construction project was such a success and the operational (particularly baggage issues) such a spectacular failure will be raked over for years to come. But this should not detract from the avoidance of all disputes on the substantive construction project.

A much more comprehensive case study is made in D Lowe (2011) *Commercial Management in Project Based Organisations*, Oxford: Wiley Blackwell.

4 Negotiation

Pink ribbon

While there is an inordinate amount of literature on the practice of negotiation, there are few theories of negotiation. The theory that does exist flows from a famous book produced from research at Harvard University, *Getting to Yes*.[1] The theories are mainly principled negotiation and positional negotiation, *Getting to Yes* says: principled = good, positional = bad. It is useful to think in terms of a continuum. Negotiation is a continuum with positional negotiation and principled negotiation at either ends.

If thought about in terms of the stages of conflict management (and dispute avoidance) and dispute resolution, then negotiation can take place at any time and might be described as very common or ubiquitous.

This is an area plagued by acronyms and buzzwords, the most common being BATNA (best alternative to a negotiated agreement) and WATNA (worst alternative to a negotiated agreement).

Introduction

This chapter introduces some of the theories of negotiation, mainly principled versus positional negotiation. Once again, it is useful to think in terms of a continuum with positional negotiation at one end, and principled negotiation at the other.

A character in Molière's *Le Bourgeois Gentilhomme*, Monsieur Jourdain, was delighted to learn that he had been speaking prose all his life – he thought that prose was something special. Equally, most people negotiate throughout their lives without realising that they are doing so, and without any training.

There are three commonly perceived attributes that most men claim to do, and do well, without any training:

- Drive.
- Make love.
- Negotiate.

It appears strange that most countries require men to pass a test only to be allowed to drive. Love making or negotiation need no training and no licence!

This chapter considers some aspects of negotiation, and unfortunately, or predictably, negotiation suffers the same fashions and fads as many management areas. Be wary of the fads and fashions – or in academic terms, demonstrate organised scepticism. A glossary is provided.

The internet revolution has, of course, affected negotiation, and negotiation via a variety of platforms is available. For a useful starting point for web-based negotiation, try any search engine.

Glossary of negotiation terms

Like any theory or practice, negotiation uses many terms in its own way. Perhaps because there is so much literature in the field, and so little theory, negotiation suffers, or enjoys (depending on your viewpoint), many, many terms. It would be impossible to cover them all, but this list is a starting point. Every book on negotiation, and there are many, will introduce its own terms.

Aardvark negotiation Negotiation between two aardvarks – always the first in any alphabetical list.

Anchoring and adjustment From neuro-linguistic programming, anchoring and adjustment is a psychological rule of thumb (heuristic) that influences the way people intuitively assess probabilities. In negotiation it is used to describe an opening position, from which a negotiator incrementally moves away from (by gains or losses) during a negotiation.

Agenda A plan for how a negotiation will progress. It can be formal and obvious, or informal and subtle. A negotiation agenda can be used to control the negotiation meeting.

Aspiration point Optimal settlement point that a negotiator hopes to achieve – the target.

Bargaining zone The gap between the respective resistance points of each party.

BATNA Best alternative to a negotiated agreement – a back-up plan which is a key feature of principled negotiation from *Getting to Yes*.

Cherry picking Generally the act of pointing at individual cases or data that seem to confirm a particular position, while ignoring a significant portion of related cases or data that may contradict that position. In negotiation it is taken as the act of picking the items which suit and maintaining that these do not affect any of the other items up for negotiation.

Consistency principle The need to appear consistent in beliefs, feelings and behaviours; a negotiator's strong psychological need to be consistent with prior act and statement.

Culture Culture is often used as a term to describe so much. Like an elephant, it is hard to define and describe, but you know it when you see it. You must understand the other party's culture.

Distributive negotiation A negotiation technique and/or type that seeks to gain at the opponent's loss. Any situation in which one person's gain is exactly equal to the opponent's loss is considered distributive. *Getting to Yes* makes the analogy of a pie and in the distributive approach each negotiator is battling for the largest possible piece of the pie.

Dyadic negotiation Two-party negotiation. A negotiation between two persons, as opposed to negotiations in which more parties are involved (also called multi-party negotiation).

Expanding the pie Finding resources to include in a negotiation that fulfil both party's needs.

Golden bridge A strategy by which a negotiator makes his or her opponent's positive decision as easy as possible. This tactic comes from *Getting Past No.*[2] Build your opponent a 'golden bridge' to retreat across.

Good guy/bad guy A tactic of team negotiation where one member of the team acts as a 'bad guy' by using anger and threats. The other negotiator acts as a 'good guy' by being considerate and understanding. The good guy blames the bad guy for all the difficulties while trying to get concessions and agreement from the opponent.

Intimidation This can take many forms: physical appearance, environmental, use of outside experts or legal authorities, use of hostages, status, threats.

Kinesics The study of movements, including posture. The most common example is body language. This area is often further complicated by culture, e.g Westeners negotiating with Easteners.

Inaction anxiety Self-imposed pressure to achieve an agreement at any cost. Often leads a negotiator to strike a deal rather than walk away and choose instead the BATNA.

Integrative negotiation A negotiation technique and/or type that seeks to expand the pie, finding win–win settlements. In an integrative negotiation, one person's gain is not necessarily another person's loss.

Interests The concerns underlying a position.

Issues Negotiable items that will be included in the formal agreement.

Lateral thinking Thinking through indirect and creative channels using reasoning and logic that is not immediately obvious and involving ideas that may not be obtainable by using only traditional steps. The term was coined by Edward de Bono.[3]

Limited authority A negotiating tactic whereby a negotiator claims to be unable to make a decision and must resort to a higher authority.

Linkage effect When one deal point of a negotiation is attached to another.

Multi-party negotiation More than two party negotiation. A negotiation between more than two persons, as opposed to negotiations in which two parties are involved (also called dyadic negotiation).

Negotiating roles Different people in a negotiating team can have different roles, such as primary negotiator, kinesics and paralanguage expert, etc.

Negotiation window The gap between the respective resistance points of each party.

NLP Neuro-linguistic programming – an approach to psychotherapy and organisational change based on models of interpersonal communication, and a system of alternative therapy which seeks to educate people in self-awareness and effective communication, and to change their patterns of mental and emotional behaviour. It was founded by Richard Bandler and John Grinder. It is not without controversy.

Non-verbal cues Body language that gives away how people are feeling and what they are thinking. This is closely related to kinesics.

Package An offer which has many elements.

Paralanguage Variations in speech: pitch, loudness, tempo, tone, duration, laughing, crying, i.e. how things are said.

PATNA Probable alternative to a negotiated agreement – a key feature of principled negotiation from *Getting to Yes*.

Position A statement of what a person/party wants in a negotiation.

Positional negotiation Positional negotiation strategy is, essentially, a manipulative approach designed to intimidate the other party to lose confidence in their own case and to accept demands.

Principled negotiation This grew from alternatives to positional negotiation. *Getting to Yes* sets out a concept of principled negotiation, with the main points being:

- Separate the people from the problem.
- Focus on interests, not positions.
- Invent options for mutual gain.
- Select from among options by using objective criteria.

Reciprocity principle Occurs when a negotiating party feels obligated to return in kind what the other side has offered or given them. This principle might result in one side making a concession because the other side has done the same.

Resistance point (RP) The point beyond which a person/party will not go. The lower limit of the range of acceptable negotiation outcomes (also called the bottom line).

Salami tactics A divide and conquer process of threats and alliances used to overcome opposition.

Strategy and tactics From military theory put shortly tactics might win the battle but strategy wins the war.

Team negotiation Different people in a negotiating team can have different roles, such as good guy/bad guy, primary negotiator, expert, e.g. kinesics and paralanguage expert, etc.

WATNA Worst alternative to a negotiated agreement – a key feature of principled negotiation from *Getting to Yes*.

Winner's curse from game theory The winner's curse is a phenomenon where, in short, the winner will tend to overpay. The winner may overpay or be 'cursed' in one of two ways: they pay more than the value of the asset such that the winner is worse off in absolute terms, OR the value of the asset is less than anticipated, so the winner may still have a net gain but will be worse off than anticipated. In negotiation the winner' s curse occurs when the aspiration point is set too low, a deal is accepted and the negotiator wonders whether the opponent would have given a better deal.

Zorrilla negotiation Negotiation between two Spanish romantic poet and dramatists – always the last in any alphabetical list.

Negotiation theory and skills

It is common for theorists to talk of two negotiation theories or strategic approaches to negotiation:

- Positional negotiation.
- Principled negotiation.

The terms positional and principled negotiating are not exclusive and in other reading material you may find them replaced with any of the following:

- Positional = competitive, compromise.
- Principled = interest based, cooperative, collaborative.

Also note that negotiations can be divided into two types:

- Dispute negotiation, focused on resolving past facts.
- Transaction negotiation, focused on reaching agreement for the future.

While it is often helpful to appreciate this difference between dispute negotiation and transaction negotiation, it is also beneficial to appreciate that many negotiation situations involve the resolution of both past issues as well as planning future relations. The theories and strategic approaches are generic and can be applied to either disputes or transactions. Mediation often involves past dispute negotiation linked to future transactions.

Distinguish strategic approach from personality

There may be some correlation between negotiation approaches and personality style, but the two do not necessarily go together. A positional negotiator may be very *pleasant* to work with in terms of demeanour, but can utilise extremely competitive tactics. Negotiator's pleasantries may themselves be part of an overall manipulative approach. A principled negotiator may be rather difficult or awkward in terms of personality, yet effectively utilise interest-based, problem-solving strategies in negotiation.

It is often argued that the most effective negotiators will have a wide array of negotiation skills, both positional (competitive) and principled (problem solving), and will effectively mix and match these approaches depending upon what the negotiator believes will work best with a particular 'negotiating partner', the specific issue being negotiated and the nature of the overall negotiating relationship (one-time transaction or continuing relations). This approach might be likened to the contingency approach to mediation described in Chapter 5.

Another view of negotiation is that certain strategies and behaviours are intended to create value (integrative and principled approaches), whereas other strategies and behaviours are intended to claim value (principled and competitive approaches).

The positional approach

Positional negotiation strategy is, essentially, a manipulative approach designed to intimidate the other party such that they lose confidence in their own case and are pressurised to accept the other side's demands. This approach is characterised by the following:

- High opening demands.
- Threats, tension and pressure.
- Stretching the facts.
- Sticking to positions.
- Being tight lipped.
- Desire to outdo, outmanoeuvre the other side.
- Desire for clear victory.

When a positional negotiator is asked how they will know that they have reached a good agreement, they might reply that the agreement is *better than fair*.

What is positional negotiating?

A positional approach involves adopting a position and aiming to negotiate an agreement while remaining as close to that position as possible. Most people are familiar with positional negotiating but it allows for only limited and fairly predictable negotiating. Negotiators adopting a positional style will assume that only one party can emerge from the negotiation a clear winner. This is often termed the 'win–lose approach'. Positional negotiating is characterised by:

- Extreme opening positions.
- Emphasis on rights.
- Aggression.
- Predictable negotiating positions.

Assumptions of the positional approach

There are certain assumptions that lie behind the positional approach to negotiation. This 'distributive' world view includes the following assumptions:

- Negotiation is the division of limited resources.
- One side's gain is the other side's loss.
- A deal today will not materially affect choices available tomorrow.

Risks of the positional approach

While positional negotiation tactics are often effective in 'claiming' already defined value, there are also certain risks. Foremost among these risks are damage to the negotiating relationship and a lessened overall likelihood of reaching agreement. The disadvantages of the positional style include:

- Confrontation leads to rigidity.
- There is limited analysis of merits of dispute and relevant criteria for resolving issues.
- There is limited development of solution alternatives.
- Difficulty in predicting the outcome of the competitive approach or control the process.
- Competitors are generally blind to joint gains.
- Competitors threaten their future relations.
- Competitors are more likely to have impasse and increased costs.

The integrative approach and the Harvard Negotiation Project

Before considering principled negotiation it is worth examining integrative negotiation and the Harvard Negotiation Project (HNP), since principled negotiation is a result of both integrative negotiation and the (HNP).

The integrative, collaborative or problem-solving approach to negotiation has been described as enlightened self-interest, rather than the egocentric variety. This approach consists of joint problem-solving, where gains are not necessarily viewed as at the expense of the other party.

Assumptions of the integrative approach

There is a different view behind the integrative approach to negotiation. The primary assumptions of the integrative approach are the following:

- Some common interests exist between parties.
- Negotiation is benefited by a full discussion of each participant's perspective and interests.
- We live in an integrated and complex world and our problems can be best resolved through application of our best intelligence and creativity.

Risks of the integrative approach

Risks of the integrative approach are based upon the common sense observation that it takes two to collaborate. If one party is unwilling to participate in integrative, problem-solving negotiation, the more collaborative negotiator may put themselves at risk in the following ways:

- The negotiator will be forced to either 'give in' or adopt a competitive stance.
- The negotiator may see themselves as a failure if they do not reach agreement.
- The negotiator lays themselves open by honestly disclosing information that is not reciprocated.

The Harvard Negotiation Project

The Harvard Negotiation Project's mission is to improve the theory, teaching and practice of negotiation and dispute resolution, so that people can deal more constructively with conflicts ranging from the interpersonal to the international.

The project, or HNP as it is commonly known, was created in 1979 and was one of the founding organisations of the program on negotiation consortium. The work of HNP routinely moves back and forth between the worlds of theory and practice to develop ideas that practitioners find useful and scholars sound. In general, HNP's work can be grouped into four categories:

1. Theory building.
2. Education and training.
3. Real-world intervention.
4. Written materials for practitioners.

A sampling of HNP activities in two categories follows.

Theory building

HNP is perhaps best known for the development of the theory of principled negotiation, as presented in *Getting to Yes*. *Getting to Yes* shows negotiators how to separate relationship issues from substance, and deal with the latter by focusing on interests, not positions; inventing options for mutual gain; and using independent standards of fairness to avoid a bitter contest of will.

Real-world intervention

HNP tested its theories in practice, often in the heat of some of the world's most intransigent conflicts. From South Africa to Latin America, the Middle East to the Balkans, HNP worked with individuals and governments on initiatives ranging from injecting a single idea at a crucial time to initiating and framing an entire process for dealing with a conflict.

A technique called 'facilitated joint brainstorming' was tested on a dispute between Ecuador and Peru to generate new options that both sides could jointly present for agreement. The conference led to a peace initiative that ultimately settled a highly contentious border dispute that had persisted for 50 years and resulted in numerous armed conflicts.

Of course, you can find more at the HNP website (http://www.pon.harvard.edu/category/research_projects/harvard-negotiation-project).

Principled negotiation

Principled negotiation was a product of HNP and grew from the alternatives to positional bargaining offered by the integrative approach. In their book, *Getting to Yes*, Fisher and Ury set forth their concept of principled negotiation. A brief summary of the main points of principled negotiation includes:

- Separate the people from the problem.
- Focus on interests, not positions.
- Invent options for mutual gain.
- Select from among options by using objective criteria.

Separate the people from the problem

Fisher and Ury suggest that we are all people first and that there are always substantive and relational issues in negotiation and mediation. They describe means of dealing with relational issues, including considering each party's perception (for example, by reversing roles), seeking to make negotiation proposals consistent with the other party's interests, making emotions explicit and legitimate, and through active listening.

Focus on interests, not positions

Positions can be thought of as one-dimensional points in a space of infinite possible solutions. Positions are symbolic representations of a participant's underlying interests. To find out interests, you might ask questions such as: 'What is motivating you here?' 'What are you trying to satisfy?' or 'What would you like to accomplish?' You may also ask: 'If you had what you are asking for (your position), what would that experientially get you – what interests would that satisfy?'

In negotiation, there are multiple, shared, compatible and conflicting interests. Identifying shared and compatible interests as 'common ground' or 'points of agreement' is helpful in establishing a foundation for additional negotiation discussions. Principles can often be extrapolated from points of agreement to resolve other issues. Also note that focusing on interests tends to direct the discussion to the present and future, and away from the difficulties of the past. If we have learned anything about the past, it is that we cannot change it. The past might help us to identify problems needing solution, but, other than that, it does not tend to yield the best solutions for the future.

Invent options for mutual gain

Before seeking to reach agreement on solutions for the future, Fisher and Ury suggest that multiple solution options be developed prior to evaluation of those options. The typical way of doing this is brainstorming, where the parties, with or without the mediator's participation, generate many possible solutions before deciding which of those best fulfil the parties' joint interests. In developing options, parties look for mutual gains.

Select from options by using objective criteria

Using objective criteria (standards independent of the will of any party) is where the label principled negotiation comes from. Fisher and Ury suggest that solution selection be done according to concepts, standards or principles that the parties believe in and are not under the control of any single party. Fisher and Ury recommend that selections be based upon such objective criteria as precedent, tradition, a course of dealing, outside recommendations or the flip of a coin.

The advantages of principled negotiating

The main advantages of principled negotiation are that it:

- Maintains relationships.
- Achieves satisfactory/efficient agreements.
- Is flexible.
- Can redress power imbalances.

Principled negotiation can be seen unlikely to provoke the aggression that might be shown in positional negotiating; the emphasis is on mutually beneficial agreements and there is no need for undue competitiveness. Parties can negotiate in an atmosphere conducive to ending on good terms with each other.

Principled negotiating can achieve satisfactory and efficient agreements since the parties are not limited to the narrow confines of rights-based arguments. The agreement can meet as many of the parties' needs as they have been prepared to reveal.

The nature of principled negotiating means that there can be flexibility over what and who is included in the negotiation.

Where a significant power imbalance exists, principled negotiating can establish that interests of the powerful party make their dependence upon the less powerful party surprisingly strong.

The disadvantages of principled negotiating

Of course there exists potential for disadvantage. Principled negotiating can take time to reach a settlement. The ultimate result should justify the time spent, but parties engaged in principled negotiating may need to be prepared to exercise patience. Principled negotiating is often far more complex than its positional counterpart. More effort will be required of parties both in preparing for and during the negotiation. It may take a number of attempts before people feel comfortable negotiating in this way, especially for those people who regard themselves as having a good track record with their positional bargaining approach. Since it can take longer and is likely to demand, overall, more labour hours, it may be seen as being the more expensive option. However, the cost should be looked at in the context of the whole negotiation, and we have seen that the end result should be a much better agreement than would otherwise have been achieved.

Some issues in principled negotiating

What if the other party is more powerful? Developing a BATNA

In the event that the other party has some negotiating advantage, Fisher and Ury suggest that the answer is to improve the quality of your 'best alternative to a negotiated agreement' (your BATNA). For example, if you are negotiating for a job and want to make a case for a higher wage, you improve your negotiating power by having another job offer available, or at least as a possibility.

What if they won't play or use dirty tricks

Fisher and Ury's answer to the resistant competitive negotiator is to 'insist' on principled negotiation in a way that is most acceptable to the competitor. The principled negotiator might ask about the competitor's concerns, show s/he understands these concerns, and, in return, ask the competitor to recognise all concerns. Following the exploration of all interests, Fisher and Ury suggest inducing the competitive negotiator to brainstorm options and to think in terms of objective criteria for decision making.

Another way of thinking about encouraging principled or integrative bargaining is to think in terms of matching, pacing, leading and modelling. To get a negotiator to shift orientations, it is critical that they first experience themselves as fully heard in terms of content, intensity and emotion. By so matching and pacing with a negotiator (asking a few clarifying questions), the negotiator will become more open to your lead and modelling of productive means of negotiating.

Converting positions to interests to positive intentions

Negotiating parties tend to come to negotiation with well-rehearsed positional statements about the truth of the situation. As wise negotiators, we know that we want to assist all parties to get below their positions to achieve a full understanding of their respective interests. If you view negotiating parties as, essentially, survivors, wanting to improve their situations, you may be able to assist negotiating parties to recognise that even the most difficult interests, such as revenge and anger, can be understood in terms of positive intentions, such as a desire for acknowledgment and respect. So, reframed, the negotiation effort can become a joint search for mutually acceptable solutions to the parties' identified positive intentions. This reframing of the entire mediation effort can dramatically shift the entire atmosphere of your negotiation.

Some negotiation basics and negotiation tactics

A bit like the three things that affect property values (location, location and location), some argue that three things are important in negotiation: preparation, preparation and preparation. Other important areas in negotiation and negotiation tactics include:

- Negotiation power.
- Opening offers.
- Stages of negotiation.

Preparation

To aid preparation, a framework might be drawn up, including:

- The causes of the dispute.
  List the causes of the dispute: is it focused on resolving past facts, or is it a transaction negotiation, focused on reaching agreement for the future?

- The range of interventions which might be helpful.
  List the interventions which might be helpful for a successful negotiation: tactics and strategy.
- The parties' needs, concerns and goals.
  Describe each party's needs, concerns and goals, and which of these need urgent attention. Rank the needs, concerns and goals, e.g. from 'vital' to 'desirable'. Describe any needs, concerns and goals that are shared, independent or in conflict. Prepare a WATNA and BATNA for each party.
- What are the facts associated with the dispute?
  Describe the facts and the evidence supporting the facts. List any facts that are agreed upon. Discuss the degree of clarity on agreed, disputed and missing facts that is necessary for advice/negotiations to begin.
- Rules and objective criteria.
  Evaluate the range of rules and objective criteria which may apply to this situation, and list the pros and cons.
- Outcomes.
  Describe the range of outcomes that are possible. Would any alternative logic help, e.g. lateral thinking? Discuss the client's targeted or preferred outcome. What outcomes will be resisted (the resistance point)?
- Dynamics.
  Consider who should:

  - engage in preliminary meetings (e.g. parties, experts)?
  - be present at the negotiation, who should not be present?

  Consider what is known about the preferred negotiation style of all parties involved. Check what authority to settle each party has. What influential people exist in the background? Consider the past patterns of interaction: are there any fears exist about a negotiation meeting? List the documents that need to be prepared and submitted. By whom should these be read and by what deadlines? Consider to what extent complex alleged facts, evidence, arguments pro and con, precedents, interests and needs and agreements need to be summarised visually.

Negotiation power

Negotiation power, simply put, might the ability of the negotiator to influence the behaviour of another: to get what you want. There are endless theories about power in negotiation. Commentators discuss a variety of aspects and qualities of negotiation power. But you have power over another to the extent that you can get another to do something that the other would not otherwise do.

Opening offers

Two vital and interrelated questions for practising negotiators and for students seeking to understand, systematise and measure negotiator behaviour are: who should make the first offer, and what form should that offer take?

Who should make the first offer?

Many negotiators try to avoid making the first offer, or begin with an exaggerated or ludicrous offer or they respond to offers with an exaggerated response. This is a natural feature of positional negotiation and is greatly alleviated by principled negotiation. Just like real men don't eat quiche, there is a feeling that real negotiators don't make the first offer. Using the techniques of principled negotiation reminds negotiators that what matters in the principle rather than the position taken on first offer, or indeed who makes that offer. In Chapter 10, game theory offers a solution to this perennial problem: making the first move gives a party the opportunity to frame the negotiation and establish precedence.

What form should the first offer take?

There are three classic ways to open negotiations:

1 Soft–high (the maximalist opening).

2 Firm reasonable (the 'equitable' opening).

3 Problem solving.

Each opening has a number of predictable and well-documented advantages and disadvantages. It is essential that skilled negotiators:

- Know how to open by any one of these methods.
- Are able to use all three openings with confidence, even though they may have a preferred style.
- Practise in damage-free simulations using alternative methods.
- Negotiate with the other side, before the first offers are made, about which of the three forms of opening is most appropriate. (This may require considerable education of the other side.)
- Are able to openly articulate the well-known advantages and disadvantages of each form of opening.

- Openly, or by known coded messages, identify to the other side which of the three openings appears to have been used. For example, negotiators use a number of codes to indicate a soft–high opening:
  - 'on the current facts, our client would be prepared to settle for ...'
  - 'our client is claiming ...'

Stages of negotiation

Like everything associated with negotiation, there are many stages offered by many authors suggesting that there are predictable stages through which most negotiations pass. Four stages are suggested and summarised here.

Stage one: orientation and positioning

Here, working relationship is established and initial negotiating positions adopted.

Stage two: argument, compromise and search for alternative solutions

Once the orientation and positioning is established argument and persuasion can proceed as the parties search for alternative solutions and seek concession.

Stage three: emergence and crisis

Now the pressure for agreement builds, and if crisis occurs then deadlock awaits.

Stage four: agreement or final breakdown

Deadlock or basic agreement occurs and the parties can wrap up the details.

These four stages demand further thought. Are there any factors which speed up or slow down the stages? Can the crisis stage be managed, alleviated, avoided? Are there any management styles that appear to be more or less effective? The proponents of principled negotiation argue that *Getting to Yes* offers a solution.

Online negotiation

As one might expect, there are opportunities presented by the use of negotiation via the internet. There are many companies offering their services as intermediaries and their servers as space for negotiation, and these are often secure negotiation rooms and platforms. Online teaching of negotiation is also widely available – try any search engine.

Facilitated negotiation

It is obvious that the parties quite often cannot communicate between themselves. This is obvious from our social affairs, our business affairs and our political affairs. An analogy often made in mediation is the chemical one. Sometimes a chemical reaction cannot take place unless a catalyst is present, or they take place more slowly. That catalyst is not part of the process and is not altered but it needs to be there. The same is true of negotiations – sometimes a third party catalyst is required.

Facilitated negotiation uses a neutral, objective person as that catalyst in negotiation sessions to help the parties reach an agreement more quickly. This neutral person has the goal of advancing discussions by ensuring that the parties understand each other's positions and facilitating settlement strategies. During negotiation, the primary aim of a facilitator is communication rather than settlement. Thus, the facilitator encourages parties to reach a settlement on their own without the facilitator's influence and makes no judgements or suggestions on how to settle the dispute under negotiation.

Win–win negotiation: how principled negotiation helped the Columbia River salmon

Southern California Edison and Bonneville Power Administration created a joint partnership which allowed both parties to meet their objectives and help the salmon in the Columbia River.

Southern California Edison generated power from the hydroelectric plant they ran on the Columbia River. Bonneville Power Administration generated power from traditional coal- and oil-fired plants. During the summer, running the traditional plants added greatly to the smog problems in Southern California. Bonneville drew water from the river to run their turbines, which not only reduced river levels but, as a consequence, also reduced hydroelectric generation. In addition to this, young salmon swim in the river during the summer months, but the low water levels made them more vulnerable to predators and more likely become disoriented and lose their way.

After creating a partnership with Southern California Edison, Bonneville drew less water in the summer months, and hydroelectric generation increased as river levels rose, which also gave the optimum conditions for the young salmon just as they needed it. The joint partnership saw power optimised with less pollution in the summer and increased salmon viability. Collaboration brought increased efficiency and helped the environment – truly a win–win.

5 Mediation

Pink ribbon

This chapter deals with mediation – the leading ADR technique, Chapter 8 deals with the other ADR techniques. Mediation is a way of settling disputes in which a third party, known as a mediator (or a neutral), helps both sides to come to an agreement which each considers acceptable. Mediation can be 'evaluative', where the mediator gives an assessment of the legal strength of a case, or 'facilitative', where the mediator concentrates on assisting the parties to define the issues. When a mediation is successful and an agreement is reached, it is often written down and forms a legally binding contract, unless the parties state otherwise.

Many ADR providers teach a facilitative model in training courses on the basis that this is considered the most successful of the mediation techniques. Any settlement that occurs is the party's own and the analogy of chemical catalysts is often made. A catalyst helps a reaction to take place between two or more chemicals, and the catalyst is not affected or changed by the reaction. Sometimes the reaction will take place without the catalyst and the effect is to speed reactions, and sometimes the reaction will not take place without the catalyst. The analogies with mediation are obvious.

Mediation is the most widely used and accepted ADR technique. While there is no prescriptive mediation process, the typical stages in a mediation might be:

1. A brief written summary of the matter in dispute is presented in advance to the mediator.
2. The parties meet with a mediator for an initial joint meeting including, perhaps, a brief oral presentation by the parties.
3. Caucus sessions, where the mediator has private meetings with the party in turn. During the caucuses the mediator often shuttles

backwards and forwards to clarify issues and search for settlement possibilities. This process is often termed shuttle diplomacy.

4 Plenary sessions are called to either continue negotiations directly, to conclude agreement, or, where the process is unsuccessful, to conclude a mediation.

Most mediators agree to a contingency approach to mediation – that is there is no set procedure but the procedure is tailored to suit the parties and the dispute in question. Like other areas of this book, mediation is plagued by inconsistencies in definition. Some people conflate the terms mediation and conciliation, others differentiate the terms: mediation = facilitation, conciliation = evaluation.

Introduction

Any history of mediation will end up in debate about where mediation started. Was it in China in antiquity? In Anglo Saxon English courts? Or in the USA in the 1970s? Various authors have laid claim to all these origins and there must be other claims. This chapter considers mediation in the UK. In many cases the term ADR is interchanged with mediation. Of course, there are other ADR techniques and these are considered in Chapter 8, but for this chapter mediation alone is considered.

In June 1991 the Master of the Rolls delivered a speech to the London Common Law and Commercial Bar Association in which he maintained:

> ADR is a PR man's dream. It conjures up visions of a factor 'X' which will do for dispute resolution what it is said to have done for washing powder. The truth is that there is no factor 'X'. Indeed, I doubt whether there is any such thing as ADR. It is simply an umbrella term or a 'buzz' word covering any new procedure or modifications of old procedures which anyone is able to think up.

That there was a sea change in opinion among the senior members of the judiciary and the Lord Chancellor can be seen when this 1991 view is considered against The Lord Chancellor's Department Consultation Paper, *Alternative Dispute Resolution – a Discussion Paper*.

> In determining the role of ADR, it is important to consider the principles on which the reforms to the civil court procedures were based. In his interim report, Lord Woolf set out the basic principles, which should be met by a civil justice system so that it ensures access to justice. These are that the system should be just in the results it delivers,

> procedures should be fair and, together with cost, proportionate to the nature of the issues involved. The system should deal with cases with reasonable speed, be understandable to those who use it, be responsive to the needs of those who use it, and should provide as much certainty as the nature of particular cases allows. And the system should be effective, i.e. adequately resourced and organised so as to give effect to the previous principles.
>
> The Lord Chancellor believes that these principles should also be used in examining ways of dealing with disputes outside the court system.

The result of this consultation was a government pledge to use ADR.

Mediation and national laws

The use of mediation does not remove the need for the parties to know the legal system within which they are operating. The legal system is vital to their rights and obligations. It is important, therefore, that a mediator knows the legal system within which the mediator operates. This begs the question: Does the mediator need to be a lawyer? There is no easy answer to the question, but some might argue that where specific and vulnerable rights and obligations exist and need protecting, then a specialist lawyer mediator is required. The obvious example is a family dispute involving children.

In litigation and other adjudicative processes, for example arbitration and adjudication, the tribunal obtains authority from legislation – Arbitration Act 1996 and Housing, Grants Construction and Regeneration Act 1996 respectively. Mediation depends on the agreement of the parties for authority. There are initiatives where courts encourage mediation, which might be described as court annexed, and in some jurisdictions, notably the USA, mandatory schemes exist. You might check the position in your jurisdiction, as we will see later that the attitude of UK judiciary and government has been to strongly support mediation, with cost sanctions, without providing for mandatory mediation. Some argue against this support and worry about mandatory mediation by stealth.[1]

Mediation principles: mediation history and types

Leaving aside the argument about where mediation started, modern early mediation theory recognised only one type of mediation. The mediator or neutral, whilst remaining in control of the process, merely facilitated the parties' negotiation in an attempt to assist the parties to *create* their own solution. This became known as facilitative (or interest-based) mediation.

Following from facilitative mediations, some parties, and some mediators, recognised that in certain situations there would have to be consideration of parties' rights and that parties unable to reach a facilitated solution would require some help via an evaluation. This developed evaluative (or rights-based) mediation. In a similar vein, a settlement-based mediation model developed.

In the 1990s, mediation developed further the initial facilitation when mediators and theorists developed a school of transformative mediation. Here, the mediator, by empowering the parties, allows all parties or their relationships to be transformed during the mediation.

Many mediation providers teach a facilitative model of mediation in training courses for mediators, and this is supported by mediation research that the facilitative model is the most successful and robust of the mediation types. Further mediators trained and experienced in facilitative models can apply other techniques and models on an ad-hoc basis. The experience of many is that mediators trained in settlement models and evaluative models are not as able to apply facilitative models. Therefore, facilitation comes first and then the mediator may choose to proceed with other techniques. Further argument is that any settlement that occurs is the party's own and the analogy of chemical catalysts is often made. A catalyst makes a reaction take place between two or more chemicals, the catalyst is not affected or changed by the reaction. Sometimes the reaction will take place without the catalyst and the effect is to speed reaction and sometimes the reaction will not take place without the catalyst. The analogies with mediation are obvious.

Of course, there are other mediation techniques and types, for example narrative mediation, but it is suggested that facilitative (or interest-based) mediation, evaluative (or rights-based) mediation, settlement-based mediation and transformative mediation, represent the techniques used in commercial disputes. Transformative mediation might be considered as a technique but its use in commercial disputes is very limited.

Facilitative mediation (interest-based)

In facilitative mediation the mediator structures a process to assist the parties in reaching a mutually agreeable resolution. The mediator asks questions, validates and normalises parties' points of view, searches for interests underneath the positions taken by parties, and assists the parties in finding and analysing options for resolution.

The facilitative mediator does not make recommendations to the parties, give advice or opinions as to affect the outcome of the case, or predict what a tribunal would do in the case. The mediator is in charge of the process, while the parties are in charge of the outcome.

Facilitative mediators want to ensure that parties come to agreements based on information and understanding. They hold joint sessions with all parties present so that the parties can hear each other's points of view, but hold private meetings (caucuses) regularly with the parties where they explore options and test the parties' positions. Facilitative mediators seek for the parties to have the major influence on decisions made, rather than the parties' advisers, legal or otherwise.

Evaluative mediation (rights-based)

In evaluative mediation the mediator assists the parties in reaching resolution by pointing out the weaknesses of their cases and predicting what a tribunal would be likely to do. Based on the parties' rights, an evaluative mediator makes formal or informal recommendations to the parties as to the outcome of the issues. Evaluative mediators are concerned with the rights of the parties rather than needs and interests, and evaluate based on concepts of fairness. Evaluative mediators meet most often in separate meetings with the parties and their advisers, practicing 'shuttle diplomacy'. They help the parties and advisers evaluate their legal position and the costs versus the benefits of pursuing a legal resolution rather than settling in mediation. The evaluative mediator structures the process, and directly influences the outcome of mediation.

Settlement mediation

Settlement mediation (compromise mediation) takes as its main objective encouragement of incremental bargaining, towards a compromise of a central point between the parties' positional demands. Mediators employing the settlement model control both the parties and the process; the mediator seeks to determine the parties' bottom line. Then, through persuasive interventions, the mediator moves the parties off their initial positions to a compromise point.

Transformative mediation

Transformative mediation is the newest concept based on the values of 'empowerment' of each of the parties as much as possible, and 'recognition' by each of the parties of the other parties' needs, interests, values and points of view. The potential for transformative mediation is that any or all parties or their relationships may be transformed during the mediation. Transformative mediators meet with parties together, since only they can give each other 'recognition'.

In some ways, the values of transformative mediation continue and expand those of early facilitative mediation, in its interest in empowering parties and transformation. Modern transformative mediators want to continue that process by allowing and supporting the parties in mediation to determine the direction of their own process. In transformative mediation the parties structure both the process and the outcome of mediation, and the mediator follows their lead.

Arguments for and against

Proponents say facilitative and transformative mediation empower parties, help the parties take responsibility for their own disputes and the resolution of the disputes. Critics say that facilitative and transformative mediation takes too long, and too often ends without agreement. There are legitimate worries that outcomes can be contrary to standards of fairness and that mediators in these approaches cannot protect the weaker party.

Proponents of transformative mediation say that facilitative and evaluative mediators put too much pressure on clients to reach a resolution. They believe that the clients should decide whether they really want a resolution, not the mediator.

Proponents of evaluative mediation say that clients want an answer when they are unable to reach agreement, and they want to know that their answer is fair. Critics of evaluative mediation say that its popularity is due to the lawyers and advisers who choose evaluative mediation because they are familiar with the process. They believe that the clients would not choose evaluative mediation if given enough information to make a choice. They also worry that the evaluative mediator may not be correct in the evaluation of the case.

Mediators tend to feel strongly about these styles of mediation and there is a healthy and useful debate. The opinion of many is reiterated: facilitation is a robust and effective model and from that basis mediators can proceed on a contingency basis. There is no one model rather a selection of techniques which can be employed depending on the circumstances. There appear to be more concerns about evaluative and transformative mediation than facilitative mediation. Facilitative mediation appears acceptable to almost everyone, although some find it less useful or more time consuming. However, much criticism has been levelled against evaluative mediation as being coercive, top-down, heavy-handed and not impartial. Transformative mediation is criticised for being too idealistic, not focused enough, and not useful for business or court matters.

Another concern is that many lawyers and clients do not know what they may get when they end up in a mediator's office. Some people feel that mediators ought to disclose prior to clients appearing in their offices, or at least prior to their committing to mediation, which style or styles they use. Other mediators want the flexibility to decide which approach to use once they understand the needs of the particular case.

Styles versus continuum

Chapter 2 shows conflict and dispute in terms of a continuum and this is also used here to differentiate between the models of mediation. Again, the differences in mediation styles and models might be seen as more a continuum than distinct differences, from least interventionist to most interventionist.

It would seem that in general mediators are on a continuum from transformative to facilitative to evaluative mediation, but are not squarely within one camp or another.

Conciliation and mediation

As with many areas in this book there is confusion in definition – be careful as the terms mediation and conciliation are often interchanged. The widely held view now is that conciliation is evaluative mediation. Often conciliation is presented as mediation where, if the parties cannot reach agreement or settlement, then the conciliator will produce a recommendation, which the parties are then free to accept or reject. Some procedures suggest that, if the parties do not reject the recommendation within a timescale, then the recommendation becomes binding. Arguments against conciliation include concern that there will never be an agreement because the parties know that the conciliator will make a decision. There will never be any mediation because the parties know that the recommendation will come. Some mediators agree and think there can be no facilitation, others like the option of the recommendation. Figure 5.1 shows mediation and conciliation at opposite ends of a facilitation/evaluation continuum.

Special features of mediation in relation to the law

There are many features of mediation which involve aspects of the law. In terms of England and Wales, one of the most pertinent, and certainly one at the heart of mediation, is the use of private meetings between the mediator and the parties. These meetings are often known as caucuses and their central role in mediation is widely acknowledged. Caucuses and caucusing

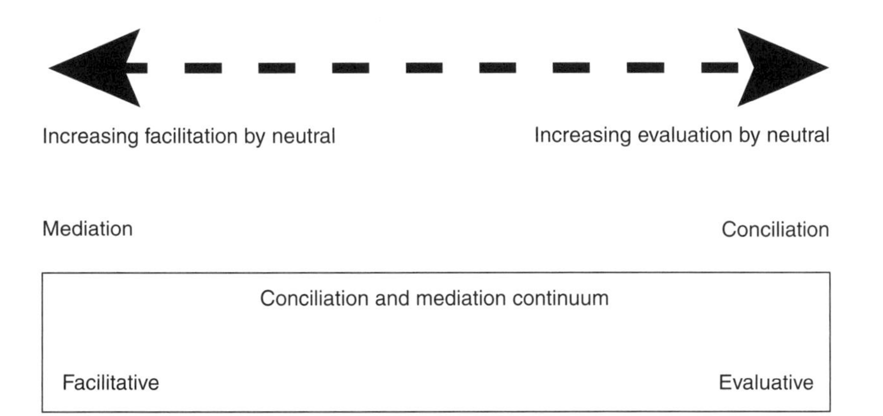

Figure 5.1 Conciliation and mediation continuum

present special problems because of *natural justice* – the principles of natural justice were derived from the Romans who believed that some natural principles of law were self-evident or natural and did not require a statutory basis. One thing is certain, that the use of caucuses by a tribunal would be a clear breach of natural justice. This separates mediation from adjudication and arbitration, and provides a powerful tool for mediators which is not available for judges, arbitrators and other tribunals.

The mediation process: civil and commercial mediation

This describes civil and commercial mediation carried out in a facilitative style, or if you like, by a facilitative mediator. Facilitative mediators are reluctant to provide decisions or recommendations, indeed some mediators will not. Any settlement that occurs is the party's own and, again, the analogy of chemical catalysts is often made.

Mediation is the most widely used and accepted ADR technique. While there is no prescriptive mediation process, the typical stages in a mediation might be: a brief written summary; parties meet with a mediator for an initial joint meeting including perhaps a brief oral presentation by the parties; caucus sessions, where the mediator has private meetings with the party in turn; and finally plenary sessions are called to either continue negotiations directly, to conclude agreement, or where the process is unsuccessful to conclude a mediation

Most facilitative mediators agree to a contingency approach to mediation – that is, there is no set procedure but the procedure is tailored to suit the parties and the dispute in question. This sometimes means that mediation is conducted without joint meetings and the mediators play a variety of roles. The mediator might act as a mere facilitator, there purely to assist communications. Alternatively the mediator acts as a deal maker, to assist the parties in finding overlap in their bargaining positions or encouraging concession and compromise. Perhaps the mediator acts more as a problem solver, assisting the parties in designing and searching for creative solutions. The mediator might act as transformer, transforming the dispute by allowing the parties a fresh insight into the issues and their positions. The final role of the mediator may be as an reality tester, to make the parties think about the merits of their cases on a legal, technical or even common sense standpoint.

FAQs

How does mediation work? Mediation involves a third party neutral taking the role of facilitator, evaluator or transformer in helping parties to reach a settlement to a dispute

What is the process at a mediation? There is no set procedure, most mediators operate a contingency approach. But the typical stages might be: a brief written summary of the dispute submitted in advance; initial joint meeting(s); private or caucuses; further joint meetings or plenary sessions.

What is the role of the mediator? Mediators play a variety of roles:

Facilitator – assisting communications and negotiations.

Dealmaker – assisting the parties by comparing bargaining positions or outlining concessions.

Problem solver – assisting the parties by suggesting creative alternatives, options and solutions.

Transformer – transforming the dispute by allowing the parties to develop a fresh insight into the issues and their positions.

Reality tester – the mediator helps the parties to consider their position: legal, technical or commercial.

Does the mediator produce an enforceable award? Facilitative mediators consider that any settlement is the parties' own settlement, they do not produce awards. The process is thought to be non-binding, consensual and non-adjudicative. It is straightforward to any agreement reached become a valid enforceable settlement agreement.

Is it enforceable? Mediation works because the presence of an independent third party neutral helps all parties concentrate on reaching a settlement. Once a settlement is agreed it can be converted into an enforceable contract, like any other settlement agreement. Mediators can help in the process of formalising in the settlement.

How are mediators trained? Formal training for mediators has historically been a two-stage process. The first stage is an academic treatment of mediation practice and procedure, followed by assessment during role-play and scenarios. After successful completion of both stages, many mediators then go on to undertake a pupillage, where they take part in a number of mediations under the supervision of an experienced mediator. The pupil mediators are again assessed by the experienced mediators for entry on to a panel of mediators when they can then take referrals in their own right.

Who are the mediators? One school of thought says that good mediators are trained in dispute resolution and therefore need not necessarily know or understand the details of the dispute. Proponents of this school say what matters is that mediators know how to resolve disputes. In the UK, most mediators are drawn from a broad range of disciplines and will have received special training in mediation. Mediators can therefore bring with them the various experience of accountancy, architecture, construction, the law, surveying, etc. to help with disputes.

What about confidentiality? It is normal for mediation providers and the mediators to require the parties and the mediator to sign a confidentiality agreement. This ensures the process is treated as a without prejudice negotiation on a strictly confidential basis. The mediator cannot be called afterwards to give evidence of what took place. Where parties are particularly concerned about confidentiality in joint meetings, the mediation can be conducted by means of separate meetings between the mediator and each party. This is such an important issue and there have been attempts in many countries to bring mediation information into court proceedings that the EU dealt with confidentiality in its Directive on Mediation (2008/52/EC)[2] so that submissions made during a mediation cannot be used in subsequent judicial proceedings if the mediation fails.

An example of the confidentiality clause common in mediation agreements might read:

> The mediation is to be conducted on a confidential basis and on a without prejudice basis unless and until, and to the extent that, the parties otherwise jointly agree.

Is proposing mediation an indication of negotiating weakness? Proposing mediation is not a sign of weakness but a suggestion of confidence on the proposing party's ability to put their cards on the table and negotiate a positive commercial solution. Skilled facilitators can help the initiation of mediation; parties are often unfamiliar with this.

When is the best time to suggest mediation? As there is no single process of mediation, there is no single correct time in the dispute to suggest mediation. However, the earlier the process can begin the more likely the parties are to achieve savings in costs and time, and notably the less likely is that the parties will have become entrenched in their positions and into the adversity of dispute. The following points are pertinent when considering suggesting mediation:

- A mediator may be able to help the parties in managing the information gathering and case evaluation process.
- Mediation does not ordinarily involve evidence presentation by the examination or cross-examination of witnesses, but takes an overall commercial approach given the need to produce a settlement.
- Mediation can be used not only for the initial attempt at settlement but it can be tried again later if and when circumstances change.
- There are few risks in attempting mediation and the parties can always abandon the process if it is unproductive.

- Most mediations will only last a short time a day or less to assess settlement possibilities.

How much does mediation cost? Of course, this is impossible to answer a bit like 'How long is a piece of string?' The service of trained experience mediators can be secured at reasonable commercial costs. It is normal that the parties bear their own costs of the mediation proceedings and share the fees of the mediator. There are often few additional costs in preparing for mediation, and mediation can assist in the information gathering process so that any costs are marginal. When compared with the costs of court proceedings, mediation cost is often negligible; it requires little additional work and often any work undertaken is useful in any other proceedings.

The benefits of mediation: the six Cs

The benefits of mediation have been summarised under the six Cs:

1 Consensual – parties seek commercial solutions assisted by their advisers and a neutral mediator.

2 Control – the parties retain control. They agree a timetable, procedure and the agenda. The outcome can be a contractual agreement.

3 Cost savings – the emphasis is on key commercial issues not on exhausting every avenue to substantiate a case or to refute the other side's evidence.

4 Continuing commercial relations – mediation often maintains commercial relation or even brings new opportunities, the emphasis is on the communication of interests between the parties, on solving the problem and on commercial solutions.

5 Confidentiality – mediation meetings are private and confidential, and should be used to explore creative solutions and agree pragmatic settlements.

6 Creative – in mediation commercial solutions are not limited by legal rules. Current and future interests and any other aspects can be explored to achieve a solution.

Preparing for mediation

How can the parties prepare and work with a mediator? It is trite to say preparation is the key, but there is a need to prepare a case in sufficient detail to allow effective negotiation while concentrating on essentials. Mediation is not a binding adjudicative process, and there is no need to explore every eventuality in case it might be relevant. However, different skills are required to summarise the core issues. Prepare a brief written summary of the dispute and the case, summarise the core issues and any evidence. This requires inclusion of only documents which are essential to the case, if there are tricky legal or technical questions as part of the dispute perhaps agree upon an independent assessment by a third party *à la Getting to Yes*. The mediator can help in appointing the independent assessor, or by using the parties' expert reports to point the parties towards a settlement. The parties need to recognise the various roles that mediators may have to play to assist the parties:

- Facilitator – to purely assist communication.
- Deal maker – to help parties find overlap in their bargaining positions or encouraging concessions.
- Problem solver – to assist in the search for creative options.
- Transformer – to transform the dispute and open up creative avenues for settlement.
- Reality tester – to provide the parties with opportunity to consider the merits of their case.

It is important that the parties work closely with a mediator in private sessions in order to make best use of the mediator's skills.

The parties must indicate clearly which information they expect a mediator to keep confidential in discussions with the other party. The mediator will normally assume that all information provided in a caucus is confidential unless otherwise indicated. It is good practice to limit the numbers of staff and representatives attending the mediation to those decision makers and professional advisers who can contribute to the settlement of the dispute. If evidence or input from others is required, particularly experts, they can be brought in as and when necessary. Often there is a kind of rubbernecking associated with disputes – everybody wants to look. Avoid this. If the case involves many parties or complex technical and legal issues, the mediator may suggest the use of co-mediators or to assessors to assist.

The parties should be prepared to be flexible. The mediation will only reach a settlement if the parties bring a genuine willingness for new solutions and both take and give concessions, in the interests of achieving a settlement. An open mind is required and a view towards pragmatic alternatives by a mediated settlement.

The relationship between the law and mediation

There is clearly a relationship between the law and mediation. In the UK, the concept of caucusing (private meetings between the neutral and the parties) is unique to mediation in that, in the mainstream processes of arbitration and adjudication, such an approach would inevitably bring about a breach of natural justice invalidating any award or decision. The attitude of UK judiciary and government is examined and it is shown that there has been a strong recent move towards supporting ADR and therefore mediation.

The attitude of UK judiciary

It is clear that the UK courts support mediation. The principle that litigation should be a last resort has the approval of the courts; in *Frank Cowl and other* v. *Plymouth City Council* 2001,[3] Lord Woolf said: 'Insufficient attention is paid to the paramount importance of avoiding litigation whenever this is possible'.

Following the Cowl case came *Dunnett* v. *Railtrack* 2002,[4] which emphasised the cost sanctions likely to flow from a blunt refusal to consider ADR. The claimant appealed against the original judgement. At the hearing, at which permission to appeal was granted, the court told the parties that they should attempt alternative dispute resolution. The defendant simply refused to consider ADR and the matter proceeded to the hearing of the substantive appeal. The appeal was dismissed, that is the defendant won, but received no costs award because the defendant had refused an offer to use ADR. It is important here to remember that the matter of parties' costs in the UK is dealt via the so-called English rule – the loser pays the winner's costs. This, of course, has a special tactical effect on the arguments for the use of ADR as a cost-saving device.

The question was asked: Can a court compel parties who do not agree to mediate, to mediate? The answer was given in the Halsey case.[5] The judgement in this case established two important principles:

> It would be wrong to compel parties to use ADR since that would amount to an unacceptable constraint on the right of access to the court and, therefore, a violation of article 6 of the European Convention on Human Rights.
>
> *However,* cost penalties can be imposed by the courts on parties that have *unreasonably* refused to consider some form of ADR. But since the general rule is that the losing party should pay the winning party's costs (the so-called English rule), it is up to the losing party to show that the winning party was *unreasonable* in refusing to consider some other form of ADR.

Two cases in 2003 were seen by many as a high watermark in the courts' support of mediation.[6] In case there was any doubt that a successful litigant might not get the costs award they might have expected because they refused to mediate, the court of appeal has repeated the view on cost sanctions (given in *Dunnett* v. *Railtrack*) in *Rolf* v. *De Guerin.*[7]

The attitude of governments: the UK government's pledge to ADR

Not only have the courts laid down their support to mediation (ADR), but on 23 March 2001 the UK Government made a pledge that government departments will only go to court as a last resort. Instead, they will settle their legal disputes by mediation (or arbitration) whenever possible. Government departments and agencies will settle legal cases by ADR techniques in all suitable cases whenever the other side agrees. The pledge is worth repeating in full:

> Government departments and agencies make these commitments on the resolution of disputes involving them:
>
> - Alternative dispute resolution will be considered and used in all suitable cases wherever the other party accepts it.
> - In future, departments will provide appropriate clauses in their standard procurement contracts on the use of ADR techniques to settle their disputes. The precise method of settlement will be tailored to the details of individual cases.
> - Central Government will produce procurement guidance on the different options available for ADR in Government disputes and how they might be best deployed in different circumstances. This will spread best practice and ensure consistency across Government.
> - Departments will improve flexibility in reaching agreement on financial compensation, including using an independent assessment of a possible settlement figure.
> - There may be cases that are not suitable for settlement through ADR, for example cases involving intentional wrongdoing, abuse of power, public law, human rights and vexatious litigants. There will also be disputes where, for example, a legal precedent is needed to clarify the law, or where it would be contrary to the public interest to settle.
> - Government departments will put in place performance measures to monitor the effectiveness of this undertaking.

The last item is, in many ways, the most far reaching. Government departments will put in place performance measures to monitor the effectiveness of this undertaking. It means that since the pledge to ADR the UK government has also been committed to reviewing and reporting the savings made. Each year it publishes a report monitoring the effectiveness of the Government's commitment to using alternative dispute resolution across Government departments and agencies. During 2008/09 the Government reported that alternative dispute resolution was used in 314 cases, with 259 leading to settlement, saving costs estimated at £90.2 million.[8]

Support beyond the UK

On April 23 2008 the European Parliament adopted the EU Mediation Directive (2008/52/EC).[9] The directive applies only to EU cross-border disputes and not to disputes within the EU member states. It covers five broad areas:

1. Mediator training and the development of, and adherence to, a voluntary code of conduct.
2. Judicial powers to invite parties to mediate.
3. Enforceability of mediation settlement agreements, i.e. obligations on member states to ensure that mediation settlement agreements are enforceable as if they were court judgements (if all parties so consent).
4. Confidentiality of mediation, so that submissions made during a mediation cannot be used in subsequent proceedings if the mediation fails.
5. The suspension of limitation periods while parties mediate.

6 Construction adjudication

Pink ribbon

Continuing the problems with definition, this chapter is about a special type of adjudication – construction adjudication – as introduced to the UK construction industry in the 1990s. Construction adjudication was brought about by legislation The Housing Grants, Construction and Regeneration Act 1996 (HGCRA). The legislation followed many government reports over a period of more than 50 years, which concluded that the construction industry, and construction projects, performed badly in terms of quality, cost and time, i.e. they were poor quality, over budget and late. A very influential report (the Latham Report) made extensive recommendations and amongst them was the recommendation that adjudication should be the normal form of dispute resolution. HGCRA effected that in Section 108. Simply put, all construction contracts must provide for adjudication, and if a contract does not, or if the provisions do not comply with Section 108, then the *Scheme for Construction Contracts* applies.

The essential requirements of the adjudication provision are:

- To enable a party to give notice at any time of intention to refer a dispute to adjudication.
- To provide a timetable with the object of securing the appointment of the adjudicator and referral of the dispute within seven days of such notice.
- To require the adjudicator to reach a decision within 28 days of referral or such longer period as is agreed by the parties after the dispute has been referred.
- To allow the adjudicator to extend the period of 28 days by up to 14 days, with the consent of the party by whom the dispute was referred.

- To impose a duty on the adjudicator to act impartially.
- To enable the adjudicator to take the initiative in ascertaining the facts and the law.

The courts have strongly supported adjudication and will enforce adjudicator's decisions. There has been strong interest in other countries and other industries.

Construction adjudication practice and procedure

This section considers a specific dispute resolution technique – adjudication as developed by the construction industry. The 'special' case of construction is considered which brought about the 'solution' of construction adjudication.

The special case of the construction industry in the UK (and worldwide)

For many years the construction industry in the UK suffered a poor reputation. These problems have intrigued, one might say obsessed, the industry and government for 50 years. Government reports on construction are nothing new and a list (incomplete) of reports since the Second World War makes for depressing reading:

1944 Report on the management and placing of contracts (*The Simon Report*).
1949 British and American productivity council.
1950 Report of the working party on the building industry (*The Phillips Report*).
1962 Survey of the problems before the construction industries (*The Emmerson Report*).
1964 The placing and management of contracts for building and civil engineering works (*The Banwell Report*).
1965 Higgin and Jessop, communications in the building industry, Tavistock publications.
1967 A survey of the implementation of the recommendations of the committee under the chairmanship of Sir Harold Banwell on the placing and management of building and civil engineering works (*Action On The Banwell Report*).
1975 The public client and the construction industries (*The Wood Report*).
1980s Faster building for commerce/industry.
1992 The building employers confederation (BEC), *Building Towards 2001*, BEC, London.

1993 Trust and money: interim report of the joint government industry review of the procurement and contractual arrangements in the United Kingdom construction industry (*The Latham Report*).
1994 Constructing the team the final report of the joint government review of the procurement and contractual arrangements in the United Kingdom construction industry (*The Latham Report*).
1998 Rethinking construction: the report of the construction task force, DETR (*The Egan Report*).
2001 Modernising construction: the national audit office.
2002 Rethinking construction: DTLR.
2003 Accelerating change: strategic forum for construction.

The solution

The consensus of all these reports was that the construction industry produced products which were of poor quality, over budget and often late. There was a further worry, more pronounced recently (since Latham in 1993), that the industry suffered from widespread dispute.

Latham made a recommendation that adjudication should be the normal form of dispute resolution. There is some debate as to what he meant by adjudication, but at the time there was a feeling that the technique of dispute resolution, arbitration mostly, was further delaying projects. This resulted in legislation in the HGCRA 1996, which produced a statutory right for all those involved in construction contracts to refer disputes to adjudication.

The HGCRA received Royal Assent on 24 July 1994. Those parts relating to construction (Part II of the Act) commenced on 1 May 1998. The Act sets out a framework for a system of adjudication in which all construction contracts must meet minimum criteria and if they fail to do so, the *Scheme for Construction Contracts* will apply.

Statutory adjudication: the process

Under Part II of the HGCRA 1996 a party to a construction contract is unilaterally given the right to refer a dispute arising under the contract to adjudication. A construction contract is defined to include an agreement to carry out construction operations. Construction operations are defined in Section 105 to include a wide variety of general construction related work, together with a list of notable exceptions. A notable exception is a construction contract with a residential occupier. The provisions only applied where the construction contract was in writing, but in 2010 this was rectified by legislation commonly known as the Construction Act 2009, the full title is The Local Democracy, Economic Development and Construction Act. This piece of legislation received Royal Assent on

12 November 2009; however, at the time of writing it is not clear when this will become law.

Section 108 sets out the minimum requirements for an adjudication procedure – there really is no alternative to reading the Act at http://www.legislation.gov.uk/ukpga/1996/53/contents. These minimum requirements may be summarised as:

Notices A party to a construction contract must have the right to give a notice of his intention at any time to refer a particular dispute to the adjudicator.

Appointment A method of securing the appointment of an adjudicator, and furnishing him/her with details of the dispute within seven days of the notice is mandatory.

Time scales The adjudicator is required to reach a decision within 28 days of this referral. It is not possible to agree in advance of any dispute that additional time may be taken for the adjudication. There are only two exceptions to the 28 day rule. First, the adjudicator may extend the period of 28 days by a further 14 days if the party referring the dispute consents. Second, a longer period can be agreed by consent of all the parties. Such agreement can only be reached after the dispute has been referred. Of course, a referring party is unlikely to agree to an extension so the period of 28 days is the default position.

Act impartially/act inquisitorially The adjudicator is required to act impartially and the adjudicator is required to take the initiative in ascertaining facts and the law. This gives the adjudicator power to investigate the issue in whatever manner he/she deems appropriate given the short time scale available.

Binding nature The decision of the adjudicator is binding until the dispute is finally determined by legal proceedings, by arbitration or by agreement. The Act does, however, go on to say that the parties may agree to accept the decision of the adjudicator as finally determining the dispute. In practice there is evidence that the parties accept the adjudicator's decision.

Immunity The adjudicator cannot be held liable for anything done or omitted in the discharge of function as an adjudicator *unless* acting in bad faith. This protection is extended to any employee or agent of the adjudicator.

In addition to this basic framework, the Act requires that any construction contract complies with the provisions of the scheme for construction contracts.

What is construction adjudication

Construction adjudication is a way of resolving disputes in construction contracts. Section 108 of the Act provides parties to construction contracts with a right to refer disputes arising under the contract to adjudication. It sets out certain minimum procedural requirements which enable either party to a dispute to refer the matter to an independent party who is then required to make a decision within 28 days of the matter being referred. If a construction contract does not comply with these requirements, the *Scheme for Construction Contracts* applies.

Adjudication does not necessarily achieve final settlement of a dispute because either of the parties has the right to have the same dispute heard afresh in court (or where the contract specifies arbitration, in arbitration proceedings). This would be a new hearing, it is not an appeal against the adjudicator's decision.

Recent experience shows that the majority of adjudication decisions are accepted by the parties as the final result. The legislation provides that adjudication can be used at any time. For example, provided that the parties have a contractual relationship, it can be used to decide contractual disputes with designers before construction begins, it can be used to resolve contractual disputes with and between designers, contractors and subcontractors during construction, and with or between them after completion.

Once a dispute has arisen between the parties, either party may seek adjudication. The adjudicator is selected within one week and must decide the dispute within a further four weeks (subject to any agreed extension). Once the adjudicator has made his decision, the other party must comply with it: if they do not, a court hearing to compel compliance can usually be obtained within a matter of days, and courts have demonstrated support for adjudication by quickly enforcing decisions.

Adjudication is thus very quick in comparison with other methods of dispute resolution such as arbitration or litigation, and it can also be used during the contract.

You can find more on the background and introduction of adjudication at www.adjudication.gcal.ac.uk. This research initiative includes a useful review as well as useful data on the progress of adjudication. For academic researchers the lack of empirical data on dispute resolution is frustrating. Construction adjudication is the ideal case study for longitudinal research and the data is both interesting and illuminating.

Adjudication: the wider term

Adjudication has a wider meaning and is used as a generic term describing a form of dispute resolution and is to be distinguished from the dictionary usage of the word, namely to determine judicially. In keeping with the general confusion over terms, adjudication in any literature about dispute resolution may have various meanings, such as the following:

- The making of a third party determination judicially (litigation).
- Other binding quasi-judicial procedures (arbitration).
- Other binding procedures (known under various names, e.g. reference to an expert, expert determination (see Chapter 8)).
- A specific procedure called adjudication, which may or may not be binding depending upon the terms of the contract within which the relevant provisions are found.
- A process called adjudication that was incorporated in all construction contracts under the provision of the HGCRA.

Before the introduction of HGCRA construction adjudication, adjudication did exist as a creature of contract, i.e. it was provided for as a contractual dispute resolution. The process was available in various forms of construction contract used in both the UK and internationally. Following the recommendations of Sir Michael Latham in 1994, provisions for adjudication are incorporated in all construction contracts under the provision of the HGCRA 1996. Construction adjudication in the HGCRA sense must be considered in stages:

- Construction adjudication procedures adopted before Latham.
- Construction adjudication according to Latham and its implementation.
- Construction adjudication under the HGCRA 1996.

The development of construction adjudication is considered in more detail because adjudication means many differing things. The chronology of the adoption of adjudication procedures in the major UK standard form construction contracts between 1976 and 1991 was considered by Bentley (1992)[1] and is updated and tabulated in Table 6.1.

All the standard form contracts shown in Table 6.1 adopted forms of adjudication that were binding until physical completion of the works and until revised by arbitration. This is in contrast to forms in other jurisdictions where the adjudicator(s) make non-binding awards or recommendations. In the JCT family of subcontracts, adjudication was mandatory but limited to issues of disputed set-off. In the main contract forms, adjudication is of wider application, but optional.

Perhaps the most celebrated form of adjudication that has been adopted was that in the Channel Tunnel contract. Not, strictly speaking, a domestic UK contract, nonetheless the clause did fall to be considered by the English courts.[2]

It is clear that adjudication as introduced by HGCRA was not new, although many present it as such.

Table 6.1 Chronology of adjudication procedures in UK standard form contracts

Date	*Form*	*Brief description*	*Binding*
1976	JCT subcontracts	Limited to interim determination of disputed set-off	Not expressly stated
1980	NSC4/4A	As above	Stated to be binding
	DOM1	As NSC4 Amendment 9: July 1990	As NSC4
1981	DOM2	As NSC4 Amendment 7: November 1992	As NSC4
1982	ACA	Optional – adjudicator empowered to deal with any dispute	Final and binding
	ACA subcontract	Where adjudicator appointed under main contract	As ACA
1983	BPF	In similar terms to ACA form	As ACA
1984	NAM/SC	As NSC4 Amendment 7: 1997	As NSC4
1985	IN/SC	As NSC4 Amendment 5: September 1989	As NSC4
	GW/S	As NSC4 Amendment 2: January 1989	As NSC4
1987	JCT works contract	As NSC4 concerning disputed set-off	As NSC4
1988	JCT81	Optional – any dispute between employer and contractor prior to practical completion	Final and binding
1989	GC/Works/1 Edition 3	Contractor may ask for adjudication by one of employer's staff not involved in contract	Binding until completion
1993	NEC	Any dispute arising under or in connection with the contract is referred	The decision is final and binding unless and until revised by the tribunal
1996	HGCRA	Those parts relating to construction (Part II of the Act) commenced on 1 May 1998. The Act sets out a framework for a system of adjudication; all construction contracts must meet minimum criteria and if they fail the *Scheme for Construction Contracts* will apply. In practice, contracts have continued to provide their own adjudication provisions. There was a fear of the scheme.	

Adjudication terms

Adjudicator The 'real' person agreed as adjudicator (i.e. it cannot be a firm it must be an individual).

ANB Adjudicator nominating bodies – a body (not being a natural person and not being a party to the dispute) which holds itself out publicly as a body which will select an adjudicator when requested to do so by a referring party.

Decision Adjudicators make decisions, they do not hand down awards (arbitrators) or make judgement (judges).

Notice of intention to seek adjudication Any party to a construction contract (the referring party) may give written notice (the notice of adjudication) of their intention to refer any dispute arising under the contract, to adjudication.

Referral notice after the adjudicator is appointed: either by agreement; via the contract provision; or by an ANB. The referring party refers the dispute in writing (the referral notice) to the adjudicator. A referral notice shall be accompanied by copies of, or relevant extracts from, the construction contract and such other documents as the referring party intends to rely upon.

Parties Normally the referring party and the responding party. Some think responding party is to be avoided. Alternatives include claimant and respondent. Use of plaintiff and defendant has all but disappeared, and, in any event, has no place in adjudication.

Is everything subject to adjudication?

HGCRA is very broad in its scope, but there are restrictions on the ability to refer disputes to adjudication. Some of these are:

- Any construction works which are not carried out within England, Scotland or Wales.
- Construction contracts with a residential occupier which relate to works to be carried out on a dwelling.
- Statutory agreements, for example to adopt a highway or sewer contracts entered into under the private finance initiative.
- Finance agreements.
- Contracts connected with nuclear processing, power generation, water or effluent treatment, handing of chemicals, pharmaceuticals, oil, gas, steel, food and drink.

- Works being carried out on land that is being sold or let under a development agreement.

How to adjudicate

Three things are required before you adjudicate. It might seem ludicrous but first you need a dispute, which normally means one party makes a claim that the other party rejects. Without that first thing there can be no adjudication. Second, there must be an agreement to adjudicate. In construction contracts following HGCRA if there isn't an agreement to adjudicate, or if the adjudication procedure isn't compliant, the *Scheme for Construction Contracts* applies. Third, someone refers the dispute to adjudication. The rest of this section describes a construction adjudication.

1 The referring party issues a notice of adjudication – the notice of intention to seek adjudication.

   The notice of adjudication is given to every other party to the contract and sets out briefly:

   - The nature and a brief description of the dispute and of the parties involved.
   - Details of where and when the dispute has arisen.
   - The nature of the redress which is sought.
   - The names and addresses of the parties to the contract.

   This notice is important because it frames the adjudicator's powers and jurisdiction. The adjudicator's jurisdiction is governed by the notice of adjudication. If the referral notice includes matters that have not already been identified in the notice of adjudication the adjudicator has no jurisdiction to deal with them.

2 Following the notice of adjudication, the parties agree as to who shall act as adjudicator (this is the easy bit, Peter Fenn at his rack rate).

   OR

   The referring party requests the person (if any) specified in the contract to act as adjudicator. The person requested to act as adjudicator shall indicate whether or not she is willing to act within two days of receiving the request.

   OR

   If no person is named in the contract or the person has already indicated that s/he is unwilling or unable to act, and the contract provides for a specified nominating body to select a person, the referring party shall request the adjudicator nominating body named in the contract (e.g. the RICS) to select a person to act as adjudicator. The adjudicator nominating body must communicate the selection of an adjudicator to the referring party within five days of receiving a request to do so.

3 Where an adjudicator has been selected the referring party shall, not later than seven days from the date of the notice of adjudication, refer the dispute in writing (the 'referral notice') to the adjudicator.

4 The scheme gives the adjudicator powers to direct the way in which the adjudication is presented and conducted.

BUT ...

5 The adjudicator shall reach their decision not later than

- 28 days after the date of the referral notice.
- 42 days after the date of the referral notice if the referring party so consents.
- Such period exceeding 28 days after the referral notice as the parties to the dispute may, after the giving of that notice, agree.

Elsewhere, the *Scheme* provides other things, amongst these:

- With the parties' consent the adjudicator may adjudicate on more than one dispute on the same contract.
- The adjudicator may resign.
- The parties may agree to revoke the adjudicator's appointment.

Adjudication – the procedure

The adjudicator must look to the contract to see if it has any procedural requirements and if any published rules apply; if so, any procedure must comply. The usual way in which an adjudicator will establish the procedure is to write to the parties on receipt of the referral setting out the way to ascertain the facts and the law. The adjudicator may take the initiative in carrying out this ascertainment or may leave it to the parties to make submissions.

The adjudicator decides on the best way to establish the facts and the law, and the best way to proceed. They may proceed entirely on the written submissions of the parties, they may meet with the parties and may obtain information from third parties. In respect of the law the adjudicator may use their own knowledge and take advice. The adjudicator may also take advice on technical matters that lie outside their own competence.

A common procedure is to allow the responding party to make a written response to the referral and then allow the responding party to respond. A meeting might then be called with both parties to allow each party the opportunity to amplify their submissions.

The adjudicator may direct that the written submissions are limited (say 2000 words) but this may prove to be unnecessarily restrictive on the parties, particularly if the dispute is at all complex.

Adjudication – the decision

There are no specific requirements as to what an adjudicator's decision must contain. Common sense says it must describe itself as an adjudicator's decision, it must identify the parties to the dispute and the project and the nature of the contract out of which the dispute arises. It must also include the adjudicator's decisions or awards on all the matters that have been referred. It should be signed and dated. There is no requirement that it be witnessed.

Guidance from the RICS[3] says that, for completeness, the adjudicator should consider the following matters during the drafting of their decision:

- A heading naming the parties to the dispute and their addresses.
- The name of the project.
- The nature of the contract between the parties.
- The way in which the adjudicator has been appointed.
- The adjudication provisions.
- The nature of the dispute between the parties.
- The relief sought.
- The issues that the adjudicator has to decide.
- The adjudicator's decision on each issue.
- The adjudicator's reasons or reasoning if required.
- The adjudicator's determination as to the costs of the parties (if so empowered).
- The adjudicator's determination as to liability for her own fees and expenses.
- The adjudicator's signature and date.

The success of adjudication

One measure of success of adjudication is the case law which has developed and the undoubted support of the courts to the process.

Another measure can be found at the website mentioned earlier: www.adjudication.gcal.ac.uk.

In 2008 the American Society of Civil Engineers ran a special edition of its *Journal of Professional Issues in Engineering Education and Practice* dedicated to 'Adjudication: tiered and temporary binding dispute resolution in construction and engineering'.[4] This edition proved so popular (among authors in any case) that the papers ran over into a second edition. It also showed the interest in many countries and detailed adjudication in different countries and in different industries. The UK experience has prompted other countries to legislate for similar procedures, and the construction experience has prompted other industries to consider adjudication.

7 Arbitration

Pink ribbon

Commercial arbitration became popular as soon as commercial people found there were advantages in staying out of court. The biggest advantage was, some say, confidentiality. Court proceedings are public and can be reported. Arbitration is a private matter – it is often said that commercial people would not seek to wash their dirty linen in public, and they may have commercially sensitive issues that require a decision. Commercial arbitration is therefore as old as commerce. Other reasons that commercial arbitration became popular include its flexibility and finality, and the fact that commercial arbitrators have expert knowledge which judges often do not.

For a while, commercial arbitration reigned supreme in England. In construction, for example, it was difficult to find a standard form of contract which did not include arbitration as the default dispute resolution mechanism. Then commercial arbitration slipped from grace, with its detractors saying that it mimicked court proceedings, that it was too slow and too expensive. It remains a stalwart of standard forms of contract.

Arbitration in England and Wales is an example of how both legislation and the common law govern. The Arbitration Act 1996 is the most recent legislation and the common law is continually updating by cases concerning legislation. The general principles as laid down by the Act are crystal clear and straightforward:

- The object of arbitration is to obtain the fair resolution of disputes by an impartial tribunal without unnecessary delay or expense.
- The parties are free to agree how their disputes are resolved, subject only to such safeguards as are necessary in the public interest.
- Courts should not intervene.

Arbitrators decide issues put before them on the evidence adduced. In England the history is not for arbitrators to act inquisitorially (to discover things by their own investigation), although the 1996 Act does allow this. The English tradition is one of adversarialism, although this is often presented as a pejorative, where evidence is tested by a system of cross-examination.

The use of commercial arbitration in England for contracts performed in England might be classified as domestic arbitration. What, then, of commercial arbitration in England for contracts performed somewhere else? This is common and indeed a major source of fee income for both arbitrators and advisors and therefore UK plc. This arbitration is classed as international arbitration. International arbitration is prestigious, lucrative and highly sought. Cities around the world compete to be major centres of international arbitration, including London, Paris, Geneva and New York.

Domestic arbitration

Arbitration is a procedure for the settlement of disputes, under which the parties agree to be bound by the decision of an arbitrator whose decision is, in general, final and legally binding on both parties. It is governed by both statute law and the common law. The principal legislation in England and Wales is the Arbitration Act 1996. Different provisions apply in Scotland and Northern Ireland and, of course, in other countries.

As a dispute resolution procedure, arbitration is the only means of dispute resolution which is an alternative to litigation because an arbitrator's award is final, binding and can be enforced in the courts.

So, arbitration is a process, subject to statutory controls, whereby formal disputes are determined by a private tribunal of the parties' choosing. Lord Justice Sir Robert Raymond provided a definition some 250 years ago which is still considered valid today.[1]

> An arbitrator is a private extraordinary judge between party and party, chosen by their mutual consent to determine controversies between them, and arbitrators are so called because they have an arbitrary power; for if they observe the submission and keep within due bounds, their sentences are definite from which there lies no appeal.

In most domestic arbitrations there is only one arbitrator. That is not the practice in many other countries, or, for example in shipping where the practice is for three arbitrators: one proposed by each party and an umpire. The arbitrator, be it one or three, is sometimes referred to as the tribunal.

The parties, referred to as claimant and respondent, are free to choose an arbitrator by agreement. The parties will often agree on an individual in whom they both have confidence, and sometimes an arbitrator with technical knowledge and expertise is prefered. For a dispute involving the quality of something supplied, say cotton or grain, it may be useful to have a technical arbitrator experienced in say cotton or grain. Both the cotton and grain industries have historically used arbitration to settle their commercial disputes.

However, the parties are in dispute and often they can't agree on anything, let alone who their arbitrator should be. In these cases, their contract often provides for an arbitrator to be appointed, for example:

> Failing agreement within 14 days after either party has given to the other a written request to concur in the appointment of an arbitrator, a person to be appointed on the request of either party by the president of the Beano Fan Club.

In extremis, an arbitrator may be appointed by the court.[2]

An arbitrator is independent and impartial and is selected by the parties, or on their behalf by an institute, on the basis of their arbitral/technical expertise, reputation and experience in the field of activity from which the dispute stems.

Almost any dispute which can be resolved by litigation in the courts can be settled by arbitration, exceptions being matters where crimes are involved or injunctions are required (because of the powers needed for enforcement) and other matters which may result in the imposition of a fine or term of imprisonment and matrimonial matters such as divorce, custody and so on. An arbitrator's award is a private matter and cannot, therefore, be effective against anyone who is not party to the dispute. This means that arbitration cannot be used in a dispute which necessarily involves parties outside the arbitration agreement, thus an arbitration agreement cannot bind third parties.

Areas where arbitration has proved especially effective include building and civil engineering contracts, shipping, rent review clauses in commercial leases, partnership disputes, insurance, manufacturing generally, computer applications, imports and exports, process industry, general trading, commodities and the engineering industry. In addition, many consumer bodies include arbitration in their standard terms, for example ABTA[3] and the concomitant ASTI.[4]

Features of arbitration

Privacy

In some cases the whole arbitration process will be considered private and confidential, but there are differences of opinion in this respect.

Flexibility

The parties may control the manner of the proceedings having regard to the nature of the dispute and to their precise needs. The parties indicate the degree of formality or informality of the procedure, unless there are pre-ordained rules or the parties are uncertain as to the procedure to adopt in which case the arbitrator will direct an appropriate procedure. There is no need for arbitration procedures to follow those of the courts, and the parties may choose alternatives, for example documents only or expedited hearing procedures.

Expertise

The parties or a 'nominating body' may appoint an arbitrator who is an expert in the matter under dispute.

Costs

Arbitration may be less costly than litigation as the use of the expert as arbitrator can save time on explanations of a technical nature. In addition, an arbitrator will normally be able to attend the hearing at a location to suit the convenience of the parties.

The costs of an arbitration are primarily time related and will depend upon the matters in dispute, the procedure chosen by the parties and their choice of representative.

Finality

The award of the arbitrator is final and binding upon the parties. It may only be challenged in the High Court on limited grounds:

1. Lack of substantive legislation.
2. Serious irregularity.
3. Error of law arising out of an award made in the proceedings (Arbitration Act 1996 – sections 66 to 71).

Enforceability

The arbitrator's award is enforceable summarily in the courts. A court will treat the award as if it were one of its own judgements.

The law of arbitration

The question of how the law of arbitration is found receives different answers depending on which law is sought.[5]

1. Questions as to the powers of the court to enforce, support, supervise and intervene are largely governed by legislation. Some powers do exist independent of legislation, created by the common law.
2. Questions as to the way in which the powers of the court should be exercised in individual cases are governed by the common law.
3. Question as to the powers of the arbitrator are governed by the arbitration agreement, which will be construed by the courts in the light of the common law.

The history of arbitration

Arbitration has a long history beyond the commercial development after the industrial revolution. Examples are to be found in ancient Greece and Rome,[6] in China, in Arabia and even in Papal practice.[7] In the Middle Ages merchants turned to each other for arbitration for their disputes. Arbitration was given codified back-up by legislation as early as 1698. The relationship between the courts and arbitration was often troubled, but in modern times they have enjoyed a close working relationship. Some argue that without arbitrators the courts could not cope, would be swamped, and that many new judges and courts would be required.[8] But by the early 1990s it was clear that commercial arbitration was in trouble. Commentators complained bitterly that arbitration had lost all its advantages, except perhaps privacy. Arbitration was slow, expensive and arbitrator's awards were often challenged.

The problems for arbitration in England and Wales were serious enough to concern government, and the taxes raised on arbitration work are a major income to UK plc. A departmental advisory committee (DAC) was established by the Department of Trade and Industry to consider how legislation might respond. In 1985 the United Nations had produced a Model Law on international commercial arbitration (UNCITRAL).[9] The DAC recommended that commercial arbitration should be fundamentally reformed by new legislation, but that legislation should NOT adopt the UNCITRAL Model Law.[10] There was a recommendation that there should be a new and improved Arbitration Act for England, Wales and Northern Ireland, with the following features:

- It should comprise a statement in statutory form of the more important principles in English law of arbitration, statutory and (to the extent practicable) common law.
- It should be limited to those principles whose existence and effect are uncontroversial.
- It should be set out in a logical order, and expressed in language which is sufficiently clear and free from technicalities to be readily comprehensible to the layman.

- It should, in general, apply to domestic and international arbitrations alike, although there may have to be exceptions to take account of treaty obligations.
- It should not be limited to the subject matter of the Model Law.

The result was The Arbitration Act 1996.[11]

The Arbitration Act 1996

The Act might be summarised under three pillars, each of which is explained by a section of the Act:

1 The general principles.

2 The duty of the tribunal.

3 The duties of the parties.

General principles

The provisions of this part are founded on the following principles, and shall be construed accordingly:

a The objective of arbitration is to obtain the fair resolution of disputes by an impartial tribunal without unnecessary delay or expense.

b The parties should be free to agree how their disputes are resolved, subject only to such safeguards as are necessary in the public interest.

c In matters governed by this part of the Act the court should not intervene except as provided by this part.

General duties of the tribunal (arbitrator)

1 The tribunal shall:

   a Act fairly and impartially between the parties, giving each party a reasonable opportunity of putting his case and dealing with that of his opponent.

   b Adopt procedures suitable to the circumstances of the particular case, avoiding unnecessary delay or expense, so as to provide a fair means for the resolution of the matters falling to be determined.

2 The tribunal shall comply with that general duty in conducting the arbitral proceedings, in its decisions on matters of procedure and evidence and in the exercise of all other powers conferred on it.

General duties of the parties

1. The parties shall do all things necessary for the proper and expeditious conduct of the arbitral proceedings.
2. This includes:
   a. Complying without delay with any determination of the tribunal as to procedural or evidential matters, or with any order or directions of the tribunal.
   b. Where appropriate, taking without delay any necessary steps to obtain a decision of the court on a preliminary question of jurisdiction or law (see Sections 32 and 45).

The aim of the Arbitration Act

Five main objectives underlie the Act:

1. To ensure that arbitration is fair, cost effective and rapid.
2. To promote party autonomy, in other words, to respect the parties' choice.
3. To ensure that the courts' supportive powers are available at the appropriate times.
4. To ensure that the language used is user friendly and clearly accessible.
5. To follow the UNCITRAL Model Law wherever possible, but NOT to adopt it.

Provisions introduced by the 1996 Act

Part 1 of the Act attempts to restate the basic principles of arbitration.

Section 1 of the Act clearly indicates that there are three principles that permeate through the Act, these are:

- Arbitration is to be a fair resolution of disputes without undue expense or delay (the object of arbitration).
- Party autonomy (reflecting the basis of the UNCITRAL Model Law).
- That the courts will play a supportive role, rather than interventionist, in arbitration matters.

The autonomy principle reflects Article 5 of the Model Law, which provides that: 'In matters governed by this Law, no court shall intervene except where so provided in this Law'.

This is viewed as a most important inclusion in the Act, since, in the past, the courts were quick to intervene in the arbitral process, thereby tending to frustrate the choice the parties made to use arbitration rather than litigation. The result internationally was that arbitration in England was to be avoided.

Section 3 of the Act sets out a definition of the 'seat of arbitration', which is an important concept internationally. The concept of 'seat' as the juridical seat of the arbitration is known to English law but may be unfamiliar to some users of arbitration. The purpose of including this section was to make the Act more user friendly to foreign users.

Section 5 states that the Act applies to arbitration agreements that are in writing. This had always been a requirement, but the 1996 Act now seeks to give a far wider definition to the meaning 'in writing' than was previously the case. The meaning is far wider than that found in Article 7 of the Model Law, but is consistent with Article II.2 of the English text of the *Convention on the Recognition and Enforcement of Foreign Arbitral Awards* – the 'New York Convention'.[12] In addition, this section recognises the fact that technology is developing very rapidly.

Section 6 defines what is meant by an 'arbitration agreement'.

Section 7 includes an express provision to the effect that an arbitration clause is separable from the main contract in which it exists. This means that if the primary contract is invalid, does not come into force or is ineffective, then the arbitration agreement survives, i.e. it is separable.

Section 9 provides that a stay to legal proceedings can be sought in respect of a counterclaim as well as a claim. Previous legislation could be said not to cover counterclaims.

Section 12 amends the court's powers in respect of granting extensions of time for commencing arbitration proceedings. In accordance with the spirit of the Act, party autonomy is given superiority and as a result it means that any power given to the court to override the bargain that the parties have made must be fully justified.

There are three cases in which the court can intervene and grant an extension of time:

1 Where the circumstances as such were outside the reasonable contemplation of the parties when they agreed the provision in question.

2 Where the conduct of one of the parties has made it unjust to hold the other to the time limit.

3 Where the respective bargaining position of the parties was such that it would again be unfair to hold one of the parties to the time limit.

Section 24 deals with the power of the court to remove an arbitrator. The old wording of previous legislation and a reference to 'misconduct' has been replaced with wording that is largely compliant with Article 12 of the Model

Law, which specifies that a court can remove an arbitrator where there are justifiable doubts as to independence and impartiality.

Section 25 sets out the consequences arising from an arbitrator's resignation. Generally, an arbitrator cannot unilaterally resign if this conflicts with the express or implied terms of his engagement. However, in certain circumstances, an arbitrator may make application to court to be granted relief from incurring liability from his resignation. Usually the court will have the power to grant the arbitrator relief if, in all circumstances, the resignation was reasonable.

Section 29 clarifies the law relating to the immunity of arbitrators. This section therefore provides that arbitrators will be immune from action unless the act or omission by that arbitrator is shown to be in bad faith.

Section 30 makes provision for the competence of the tribunal to rule on its own jurisdiction. This is traditionally called the doctrine of 'Kompetenz-Kompetenz', which is an internationally recognised doctrine and is set out in Article 16 of the Model Law. This concept is so important that it has its own YouTube entries and a German Wikipedia page! The value of this doctrine is that it avoids delays and difficulties when a question is raised as to the jurisdiction of the tribunal. Clearly, the tribunal cannot be the final arbiter of a question of jurisdiction, but failure to include this power would result in a recalcitrant party completely disrupting the arbitral proceedings.

Section 31 sets out how a challenge to the jurisdiction of the tribunal can be made, and the circumstances in which it must be made. This section reflects most of Article 16 of the Model Law, with some small exceptions. This Section allows the tribunal to make an award as to its own jurisdiction, either as an interim award or as part of its award on the merits. Section 67 provides the mechanism for challenging the jurisdiction rulings in such awards.

Section 33 sets out the general duties of the tribunal, which is one of the central pillars of the Act and is based upon Article 18 of the Model Law. This section sets out in the simplest of terms the manner in which the tribunal should approach and deal with its task, which is to do full justice to the parties.

Section 33 not only instructs the arbitrators to 'act fairly and impartially as between the parties', but it also compels them to: 'Adopt procedures suitable to the circumstances of the particular case, avoiding unnecessary delay or expense, so as to prove a fair means for the resolution of the matters falling to be determined'.

Section 35 provides that the parties may agree to consolidate their arbitration with other arbitral proceedings or to hold concurrent hearings. Previous to this there were concerns that the failure of arbitration to allow consolidation produced inefficiencies and it is said this section protects the parties autonomy.

Section 38 significantly redefines the relationship between arbitration and the court. Wherever a power could be properly exercised by a tribunal rather than the court, provision is made to allow and facilitate this. This reduces the need to incur the expense and inconvenience of making application to court during arbitral proceedings.

Section 41(6) of the Act provides a sanction for failure to comply with the tribunal's order for security for costs, whereby the tribunal can follow the practice of the English commercial court. As a consequence, the tribunal is now empowered to make an award dismissing the claim, should one of the parties fail to provide the security the tribunal orders. This ensures that the matter is finally disposed with, not left dormant, which would be the natural result should the proceedings be stayed.

Section 39 gives the tribunal, on agreement by the parties, the power to make temporary or provisional orders regulating financial arrangements between the parties, but which will be subject to reversal when the underlying merits are finally decided by the tribunal. In the absence of agreement, the tribunal is not so empowered. It is important to note that the tribunal does not have extended powers to issue injunctions or relief that remain the sole jurisdiction of the courts.

Section 40, another of the central pillars, provides that the parties have a general duty to do all things necessary for the proper and expeditious conduct of the arbitral proceedings.

Section 41 sets out the tribunal's powers in case of a party's default.

Section 42 sets out that the court has the power to order compliance with the tribunal's peremptory orders as set out in Section 41.

Section 44 redefines the relationship between the courts and arbitration by virtue of the fact that it sets out the powers that can be exercised by the court in support of the arbitral proceedings. This provision corresponds with Article 9 of the Model Law. The powers given to the court are only those necessary when the tribunal cannot act effectively.

Section 46 provides that the parties have the freedom to decide that the tribunal will make a decision not in accordance with a recognised system of law. This is sometimes known as equity or *ex aequo et bono* clauses, that is to say, generally recognised principles of fairness and justice.

Section 47 provides that the arbitral tribunal may make awards on different issues.

Section 49 provides that the arbitral tribunal has the power to award compound interest.

Section 52 re-enacts Article 31 of the Model Law, which provides what form the award is to take. Awards *must* be reasoned unless both parties have indicated otherwise.

Section 53 provides that any award made in the proceedings of an arbitration whose seat is in England, Wales and Northern Ireland will be treated as having been made at that seat, regardless of where it was signed, dispatched or delivered.

Section 57 provides that the tribunal has the power to correct an award.

Sections 59 to 65 provide a code dealing with how the costs of an arbitration should be attributed between the parties. The question of the right of the arbitrators to fees and expenses is dealt with earlier in Section 28.

Section 59 defines costs.

Section 62 provides that, unless the parties agree otherwise, the right to claim costs extends only to the recoverable costs.

Section 63 provides that the tribunal may determine the recoverable costs of the arbitration on a basis it thinks fit. However, if the tribunal does not determine the recoverable costs of the arbitration, a party may make application to the court for it to decide upon the recoverable costs of the tribunal.

Section 64 provides that only reasonable costs and expenses as are appropriate in the circumstances are recoverable.

Section 65 gives the tribunal power to limit in advance the amount of recoverable costs. This could be used to reduce unnecessary expenditure.

Section 66 sets out the powers of the court in relation to arbitration awards.

Section 67 provides in what instances a party may seek to challenge an award based upon the jurisdiction of the tribunal. This section also provides a mechanism for challenges to the jurisdiction by someone who has taken no part in the arbitral proceedings. Finally, in order to avoid unnecessary delays, the tribunal is empowered to continue with the arbitral proceedings in the face of a court challenge to its jurisdiction.

Section 68 provides in what instances a party may seek to challenge an award based upon a serious irregularity affecting the tribunal, the proceedings or the award.

Section 69 sets out the instances upon which an appeal on a point of law is allowed.

Section 70 sets out the provisions that are to apply to an application or appeal under Section 67, 68 or 69. The time limit for lodging an application or appeal is 28 days of the date of the award, or when the applicant was notified of the result of the process.

Section 72 sets out the saving for rights of person who takes no part in the proceedings.

Section 73 provides that recalcitrant parties or those who have had an award made against them cannot seek to avoid honouring the award or delay proceedings by raising points on jurisdiction, etc. which they could, and should have, discovered and raised at an earlier stage of the proceedings. Failure to raise the necessary objection at the time it becomes apparent results in waiver of the right to object at a later stage. In this way, a recalcitrant party cannot avoid the effects of a final and binding award that goes against them. This reflects Article 4 of the Model Law but goes further by requiring a party to arbitration proceedings who has taken part, or continued to take part, without raising the objection in due time, to show that, at that stage, he neither knew nor could with reasonable diligence have discovered the grounds for his objection.

Section 74 provides that arbitral institutions are immune from suit in respect of anything done or omitted in the discharge of their function unless the act or omission is shown to have been in bad faith.

Part II of the Act sets out the difference between domestic arbitration agreements and international arbitration agreements. A differentiation is maintained because the rules for obtaining a stay of legal proceedings differ. The New York Convention applies to international arbitral awards, and domestic enforcement provisions relate to domestic awards.

Part III of the Act re-enacts the substance of the 1975 Arbitration Act which gave effect to the New York Convention. As a consequence, the 1975 Act has been repealed.

Part IV of the Act merely sets out the administrative elements to its enactment and application.

Arbitration practice and procedure

Adversarial or inquisitorial

Arbitrators decide issues put before them on the evidence adduced. In England the history is not for arbitrators to act inquisitorially (to discover things by their own investigation), although the 1996 Act does allow this under section 34(2)(g).

The adversarial system is a legal system where, via advocacy, parties' positions are put before an impartial adjudicator (an arbitrator in this case). Alternatively, the inquisitorial system has an arbitrator whose task is to investigate the case. Although many present the term adversarial as a pejorative, it is an efficient way to test the evidence.

Proving a case: it is the claimant's job to prove the case alleged; the defendant could simply deny and put the claimant to prove the case. This is a 'high risk' strategy and, although not unheard of, the defendant will normally produce evidence to deny the allegations. In civil cases the proof required is described as *on the balance of probabilities* (the American term is *the preponderance of the evidence*). In criminal proceedings the standard is *beyond a reasonable doubt*, but in civil proceedings the standard is *the balance of probabilities*. So, in civil proceedings, to prevail the party bringing the action merely has to demonstrate on the evidence that they are right on the balance of probabilities, i.e. 51 plays 49.

Two related, but very different, concepts must be differentiated: burden of proof and standard of proof.

Burden of proof

The party bringing the action, normally the claimant, has the burden of proof. The claimant must prove his case; the defendant need do nothing. This is the burden of proof and the maxim is: he who alleges must prove. There are occasions where the burden of proof shifts, normally where there is a strict liability; but for most purposes the burden of proof lies with the claimant.

Standard of proof

The party with the burden of proof must extinguish that burden to a certain degree – the standard of proof. In commercial civil disputes that standard is the balance of probabilities. Be aware that most non-lawyers are only aware of the criminal burden (beyond a reasonable doubt).

Evidence and expert evidence

The claimant proves his case by adducing evidence. There is a distinction between evidence of fact and evidence of opinion. The general rule in England and Wales is that the only evidence that can be given is evidence of fact – that is, a witness can give evidence of those things experienced by the five senses: sight, sound, touch, smell and taste. Thus, a witness can say, 'I saw the drawings,' but witness cannot say, 'the drawings were good' (or bad), because that would be an opinion. However, the law recognises that judges (and arbitrators) are not polymaths, and they require help on specialist issues. In such cases, expert evidence may be permitted.

The procedure at arbitration

The arbitration is begun by the claimant writing to the respondent, requiring them to agree on the appointment of an arbitrator (notice of arbitration). It is sensible for the parties to agree their arbitrator, but as earlier commented, they are in dispute and can often agree on nothing. It is common for the claimant to provide a list of three arbitrators for the respondent to choose from. If the parties cannot agree, it may be that the contract provides for a third party to appoint an arbitrator.

The general procedure is shown below, although modern procedure is flexible and alteration can be made.

1 Notice of arbitration.
2 Appointment of arbitrator.
3 Preliminary meeting.
4 Interlocutory stage
    - order for directions
    - reducing the issues to writing
        - statements of case (pleadings)

        - further and better particulars
        - discovery and inspection.

5 The hearing

- claimant opens
    - claimant calls evidence (fact before opinion)
    - respondent cross-examines
- respondent opens
    - respondent calls evidence (fact before opinion)
    - claimant cross-examines
    - respondent closes
- claimant closes.

6 Arbitrator considers.

7 Arbitrator determines, arbitrator's award.

8 Enforcement or appeal.

International arbitration

International arbitration is selected by many of the world's leading international companies. They insert an arbitration clause into their agreements with trading partners and opt to have disputes in connection with the contract decided by private tribunals (arbitral tribunals) rather than litigating them in national courts.

The problem of definition, which is present throughout this book, comes up again. There is no uniform definition, but many arbitrations which might be considered 'international' include the following:

1 Wholly foreign case, e.g. arbitration between the parties, one from Germany and one from China, to a project in Spain. The arbitration held in Paris, the arbitration in French, the arbitrator (or one of the tribunal) is French.

2 A foreign case, e.g. arbitration between the parties, one from Germany and one from China, to a project in Spain. The arbitration held in London, the arbitration in English, the arbitrator (or one of the tribunal) is English.

3 A foreign case, e.g. arbitration between the parties, one from England and one from China, to a project in Spain. The arbitration held in Paris, the arbitration in French, the arbitrator (or one of the tribunal) is French.

4 A foreign case, e.g. arbitration between the parties, one from England and one from China, to a project in Spain. The arbitration held in London, the arbitration in English, the arbitrator (or one of the tribunal) is English.

The most common reasons for opting for international arbitration are:

- That arbitration awards (broadly equivalent to a court judgement) are easier to enforce internationally and cannot so easily be dragged to appeal for years.
- That neither party is willing to have disputes decided by the other party's national courts.
- That the parties are keen to have their disputes resolved privately.

International arbitration allows parties to have their disputes decided by a neutral tribunal, which can be made up of legal and/or industry experts of the parties' own choosing, using procedures which they can influence. The normal procedure is for the tribunal to be composed of three arbitrators. One from each party and one neutral umpire.

Lex Mercatoria

Literally, the Law Merchant – some argue that international disputes should not be subject to any national law but to international concepts of commercial law. These concepts are collectively known as *Lex Mercatoria.*

Delocalisation or territoriality

The proponents of delocalisation argue that international commercial arbitration should be completely independent of any form of state control, except perhaps at award enforcement. The courts are not to become involved in any way with the arbitration. The delocalisationists have not found widespread acceptance.[13]

The proponents of territoriality (the generally accepted view) argue that arbitration cannot be conducted in a vacuum and some things cannot be left to the parties:

- Appointment of the tribunal (e.g. when the parties cannot agree).
- Removal of an arbitrator (e.g. for bias or other misconduct).
- Provide assistance (e.g. enforcing attendance of witness).
- Establish if an award is valid and final.

The use of international commercial arbitration

International commercial arbitration is a complex and important topic, a survey carried out by PricewaterhouseCooper produced a report *International Arbitration: Corporate Attitudes and Practices* (2008).[14] Respondents to the survey displayed a strong preference for international arbitration, as an alternative to transnational litigation, to resolve international disputes. Arbitration is perceived as a private and independent system, largely free from external interference. In certain industries, such as shipping, energy, oil and gas or insurance, international arbitration is the most commonly used dispute resolution mechanism.

8 Other ADR techniques

Pink ribbon

This chapter is a catch-all chapter which attempts to describe the ADR techniques not included elsewhere. The problem with that approach is that there are many, many techniques, some of which are very specialised. And since you started reading this book/chapter/paragraph, someone has invented a new one. And some are so specialised and speculative, for example, the Peter Fenn patented system where Peter Fenn is appointed at £500 per hour to resolve disputes by juggling and unicycling.

There is, therefore, no attempt to include all the techniques, rather, techniques are included on the grounds of application or interest. Bearing that in mind, there is incredible interest in med-arb, but almost no application. Mini-trial (executive tribunal) is similar, but with some big-ticket application examples. Early neutral evaluation has widespread interest in the USA and the UK, and widespread application in the USA but much less so in the UK. Expert determination, which is known by many names, is applied widely in some industries and carries onerous health warnings – not because it is dangerous, but parties must understand what they are agreeing to because it is agreement. Contracted or project mediation is included here rather than a special case of mediation because it requires contract or project provision. Dispute review boards, again, come under many very similar names and the acronyms can be tortuous: DRBs, DABs and DRPs among others. Finally, dispute review advisers are considered.

Introduction

There are many definitions of ADR:

- Alternative dispute resolution.
- Amicable dispute resolution.

- Appropriate dispute resolution.
- Another disappointing result.
- Another damn rip-off.

In the UK, because of the long history of arbitration, the term ADR has normally been taken to mean those techniques alternative to litigation and arbitration, i.e. arbitration is not ADR. This has caused confusion since arbitration in the USA is considered to be ADR! The water has been further muddied by the Lord Chancellor's Department Current Consultation Paper, *Alternative Dispute Resolution – a Discussion Paper*, which states: 'The phrase Alternative Dispute Resolution now covers a variety of processes that provide an alternative to litigation through the courts, and can be used to resolve disputes where those involved would be unlikely to resort to the courts'.

ADR processes include arbitration, early neutral evaluation, expert determination, mediation and conciliation. For some at least, negotiation within the processes of litigation forms part of the ADR repertoire, with important links to existing litigation practice.

Other more formal mechanisms for resolving disputes, such as the private sector ombudsman schemes, utility regulators, trade association arbitration schemes in certain trade sectors, and even tribunals, can also provide alternatives to the courts in some circumstances.

The various processes have very different characteristics. It can sometimes be unhelpful and confusing to group them together under one heading. A useful distinction is made between processes in which a neutral third party makes a decision and those where the neutral offers an opinion, and/or seeks to bring to the parties to an agreement. Here, the term *alternative adjudication* is used to encompass decision-making processes other than litigation through the courts, such as arbitration, and expert determination, ombudsmen and regulators. *Assisted settlement* is used to encompass processes designed to help the parties come to an agreement, such as mediation, conciliation and early neutral evaluation. Of course, it is possible to have hybrid processes. *Med-arb*, for example, describes a process where there is an initial agreement to mediate the dispute and, if that fails to achieve settlement, to submit outstanding issues to arbitration. In addition, some ombudsman schemes incorporate mediation into their procedures.

In this context, the word *alternative* conveys only that these are methods of dispute resolution which are not those in general use in litigation (which is why, for some, negotiation does not fall within the ADR territory). It does not imply that the use of ADR techniques is in some way second best to going to court. A case has been made for referring instead to appropriate dispute resolution, to reflect the arguments that some ADR techniques are better suited to the needs of some cases or litigants than court proceedings. The term alternative *dispute resolution* is, however, probably now so well established that there is little prospect of changing it. Semantics of this type is great fun but of limited value.

Of course, some of the procedures now considered as ADR are considered elsewhere in this book, and discussing what constitutes ADR is one of life's more meaningless and pointless activities. This chapter considers:

- Med-arb.
- Mini-trial (executive tribunal).
- Early neutral evaluation.
- Expert determination.
- Contracted or project mediation.
- Dispute review boards.
- Dispute review advisers.

The history of ADR

There exists a *widely held belief* that ADR is a recent development, and that the techniques referred to as ADR came from the USA. An interesting variant, variously quoted, is that society and culture may affect dispute resolution, the most common manifestation of this being that 'Eastern' cultures are less attracted to confrontation and therefore ADR came from the East. Evidence for the first belief is contained in just about every chapter of Fenn and Gameson (1992)[1] and the second belief is widely alluded to in the same book.

Of course, the *widely held belief* that ADR is a modern development is simplistic, but the fact that the American Bar Association (ABA) noted in 1985 a dearth of scholarship on the historical aspects of ADR[2] indicates that a revival in interest in ADR took place in the mid-1980s. This is often the date ascribed by many of the ADR advocates when recounting the *widely held belief* on the development of ADR.

From the ABA's note of the lack of research into the history of ADR came a research programme. That research considered dispute processing in Anglo-Saxon England, one of the earliest stages on English legal history. The central findings of the research are that the Anglo-Saxons used an array of dispute resolution processes. The array included processes, which might be compared with:

- Negotiation.
- Adjudication.
- Arbitration.
- Mediation.

In addition, these processes were available to the parties during the life of an action on a dispute–processing continuum. The processes and the interrelationships of the dispute–processing continuum were aimed at fostering respect for the legal processes and effecting the peaceful and enduring resolution of disputes and promoting the reconciliation of the parties.

It could be argued that the recent interest in ADR shares many of these sentiments. Perhaps not the fostering of respect for the law and legal processes since business, commerce and industry have become more sophisticated. Many would argue that commercial people have no interest in the majesty of the legal process; their concern is for the effective resolution of disputes; both peaceful and enduring. At the heart of the ADR philosophy has been the reconciliation of the parties in order that they might have continuity in their commercial affairs.

ADR is not a recent development and although it suits many people to claim truth in the *widely held belief* the history of ADR can be reliably traced to fifth-century England.

Other ADR processes

This part provides a outline of other ADR processes than civil and commercial mediation. The forms of ADR considered are:

- Med-arb.
- Mini-trial (executive tribunal).
- Early neutral evaluation.
- Expert determination.
- Dispute review boards.
- Dispute review advisers.

Med-arb

Med-arb is a process that has been much debated in recent times, where, if the mediation fails to produce settlement, the mediator might take the role of arbitrator. However, although the idea is attracting interest, there remain some fundamental matters of principle, which may affect the validity of a binding award for such a tribunal. These include the argument that such an award would be against the rules of natural justice in that a party to the process is unaware of what is being said in one of the private caucusing sessions and is therefore unable to reply. There is little evidence of the adoption of this method of dispute resolution but no doubt the debate is likely to continue for some time. Some have wondered if the reverse process, i.e. arb-med would remove the issues of principle objection the problem with natural justice.

No other ADR technique attracts the same debate as med-arb. As the name suggests, med-arb is essentially a two-stage hybrid ADR process. During the first stage the parties attempt to settle their dispute amicably using mediation, if settlement cannot be found then the parties move to the second stage of arbitration. The essential characteristics, and the one that causes all the debate, of this technique is that mediator in the first stage becomes the arbitrator for the final and binding stage.

Med-arb arises either through a contractual provision or by a party agreement once the dispute has arisen. The proponents of med-arb argue the advantage that it combines the benefits of a possible mediated settlement with the finality of arbitration.

Med-arb recognises that arbitration may not resolve all the issues between the parties but limits the arbitration solely to the intractable disputes, thereby bringing the cost and time saving to the parties.

Conversely, the detractors of med-arb express concerns over such a procedure, claiming that it compromises the neutral's capacity to act initially as facilitator and then as the arbitrator, without restricting the flow information. The fundamental objection to such an approach is that the parties will not wish to reveal confidential information during caucuses with a mediator which may then influence the arbitrators view of them during arbitration.

Notwithstanding the wide-ranging debate on med-arb, few seem to have any real experience of the technique being used.

Executive tribunal or mini-trial

These are management-based mediation-assisted processes. The term mini-trial is favoured in the USA and in the UK executive tribunal has been adopted. The procedure is one of structured settlement, combining negotiation, mediation and adjudication. Lawyers for each party are given a limited time to make a presentation on the dispute, in the presence of senior executives from each company. The executives are required to have the authority to settle the dispute, and following the presentations they hold a meeting moderated by a neutral. The neutral acts as facilitator towards a settlement and can if required offer either a binding or non-binding opinion.

The mini-trial procedure in the USA developed from a large dispute over a period of three years which had exchanged over 100,000 documents. In an attempt to settle the dispute an alternative was suggested. The alternative was developed from the observation that often disputes develop a self-fulfilling dimension within organisations, in that the time involved and committed becomes so large that the parties lose sight of the commercial reality of the dispute. Often senior executives are appalled when they see the extent of the dispute.

The major benefits proposed are:

- Senior executives become involved and realise the nature and severity of the dispute.
- Senior executives are given an opportunity to hear the arguments from both sides.
- Senior executives are able to meet and discuss settlement.
- Senior executives are not constrained by legal min/lose remedies.

Few experiences have been documented[3] and these few have been on major projects. Many are confused about the process, believing that it relates to site negotiations which have reached the point where company executives become involved.

Early neutral evaluation (ENE)

ENE is one of a number of forms of adjudicative and binding dispute resolution. These forms might be though of running from an adjudicative neutral type of conciliation through to a system of mini-trial. There has been considerable interest in ENE due to a series of court initiatives. There are few reports that the systems of ENE have been used in practice, but the evidence supports the wider contention that the parties, their advisers and the dispute resolvers – adjudicators, arbitrators, conciliators, judges and mediators – now have a continuum of techniques to use depending on the dispute in hand. A useful statement might be one applied to every dispute technique described in this book: 'ENE (insert any other technique) is simply one tool in the toolbox of dispute resolution'.

ENE and the courts[4]

A Commercial Court Practice Statement of 10 December 1993 encouraged the parties to disputes in the commercial courts to consider the use of ADR.[5] From then, the judges in the commercial court adopted a practice at the first interlocutory hearing when directions were to be given of evaluating the action. If any of the issues were appropriate then attempts would be made to settle the dispute by ADR. The judges considered if the parties had attempted ADR, and if not they invited the parties to take positive steps to set in motion ADR procedures. As a result of this, practice statement changes were made to the standard questions to be answered by the parties in preparation for the summons for directions. Similar changes were also made as part of the pre-trial checklist. Cynics suggested that solicitors merely included the necessary changes in order to comply with the letter of the practice statement, that they had no intention of complying with or entering into the spirit.

A further practice direction issued by the commercial court,[6] stated that the judges of the commercial court in conjunction with the commercial court committee had considered whether it was desirable that further steps should be taken to encourage the wider use of ADR. The practice direction proposed that, if, after discussion with the parties' representatives, an early neutral evaluation is likely to assist in the resolution of the matters in dispute, then the judge may offer such ENE himself. Alternatively, the judge could arrange that another judge carry out the evaluation. Where the ENE was to be carried out by a judge it was stated that the judge would, unless the parties agreed otherwise, take no further part in the proceedings.

Guidance notes published by the commercial court described ENE as: 'The function of this procedure is to provide the parties to a dispute with a non-binding assessment by a neutral of their respective chances of success were the litigation to be pursued'.

The initiative in the high court might at first sight seem surprising, but as Judge Toulmin points out those who are surprised will be equally surprised to discover that ENE is done as a matter of course in Israel and New Zealand. Further, the procedures in both countries are similar to the UK and reports that the system appears to work well in both jurisdictions.

The procedure in ENE will necessarily vary in each case, but for the dispute put before the Technology and Construction Court (TCC) in June 1998 the following procedure was initiated in the Case Management Conference in April 1998:

- Agreed statement of issues and chronology to be lodged at court by 1 June.
- Pleadings; and defendant's response to be lodged at court by 1 June.
- Case outline (ten pages maximum) to be lodged at court by 1 June.
- ENE to be held in court 8 June.
- Each side 30 minutes to open and respond.
- 60 minutes for judge to question parties.
- Period of reflection before judge delivers assessment.
- Nothing said at the ENE would be used in litigation or for any other purpose.
- The judge would be disqualified from any other involvement in the proceedings.
- Each party to bear their own costs.

The report of the commercial court committee acting party on ADR February 1999

The report warned that only four such ENEs had been conducted in the Commercial Court (this has been doubted by others[7]) and explained the low number by the lack of familiarity of the parties and their advisers. The report considered that ENE could only assist parties in cases where there were few distinct issues on which preliminary views could be readily expressed without substantial presentations on the merits from either party. ENE could be a useful tool for the judge to encourage or persuade settlement.

ENE application

At an early stage of the case the parties may apply for ENE, but the problems are that issues may not be sufficiently clarified, or developed, to allow ENE. ENE may be more appropriate where the parties' position on particular issues have been defined and hardened such that they are only likely to change on the authoritative view of a third party. For the neutral to be able to give a view, and for that view to carry the necessary weight, all or most of the evidence, statements, reports and documentation must be available to the neutral.

Timing of ENE is crucial. Too early and the parties have not had sufficient time to exchange and appraise the evidence and may retain feelings that their position could be improved. Too late and there will be little by way of cost savings.

Expert determination (submission to expert, reference to an expert, expert adjudication)[8]

Note that statutory construction adjudication under the *Scheme for Construction Contracts* and the HGCRA is dealt with separately in Chapter 6.

These are long-established procedures in English law and have been used across a number of industries. Examples include accountants valuing shares in limited companies, valuers fixing the price of goods, actuaries carrying out valuations for pension schemes, certifiers of liability for on-demand performance bonds, certifiers under construction contracts, and adjudicators who are said to be acting 'as expert and not as arbitrator'.

The law governing the extent to which the courts may interfere in expert determination has been reviewed in a number of recent cases. *Jones* v. *Sherwood Computer Services plc* (1992)[9] held that, where parties had agreed to be bound by the report of an expert, the report, whether or not it contained reasons for the conclusion in it, could not be challenged in the courts on the ground that mistakes had been made in its preparation unless it could be shown that the expert had departed from the instructions given to him in a material respect.

Jones was followed in *Nikko Hotels (UK)* v. *MEPC* (1991)[10] which held that, if parties agree to refer to the final and conclusive judgement of an expert, an issue that either consists of a question of construction or necessarily involves the solution of a question of construction, the expert's decision will be final and conclusive and, therefore, not open to review or treatment by the courts as a nullity on the ground that the expert's decision on construction was erroneous in law, unless it can be shown that the expert has not performed the task assigned to him. If he has answered the right question in the wrong way, his decision will be binding.

Jones was also approved in the *House of Lords in Mercury Communications Ltd* v. *Director General of Telecommunications* (1996).[11]

In *British Shipbuilders* v. *VSEL Consortium plc* (1997),[12] the principles were summarised:

Five principles govern the status of decisions of a person occupying the role of the expert:

1 Questions as to the role of the expert, the ambit of remit (or jurisdiction) and the character of remit (whether exclusive or concurrent with a like jurisdiction vested in the court) are to be determined as a matter of construction of the agreement.

2 If the agreement confers upon the expert the exclusive remit to determine a question, the jurisdiction of the court to determine that question is excluded because for the purposes of ascertaining the rights and duties of the parties under the agreement the determination of the expert alone is relevant and any determination by the court is irrelevant. It is irrelevant whether the court would have reached a different conclusion or whether the court considers that the expert's decision is wrong, for the parties have in either event agreed to abide by the decision of the expert.

3 If the expert, in making his determination, goes outside remit, e.g. by determining a different question from that remitted or in determination fails to comply with any conditions which the agreement requires compliance with in making determination, the court may intervene and set the decision aside. Such a determination by the expert as a matter of construction of the agreement is not a determination which the parties agreed should affect the rights and duties of the parties, and the court will say so.

4 The court may set aside a decision of the expert where (as in this case) the agreement so provides if determination discloses a manifest error.

5 The court, in advance of a determination by the expert, may determine questions as to the limits of remit or the conditions which the expert must comply with in making the determination but will save in exceptional circumstance decline to do so. This is because the question is hypothetical only proving necessary if, after seeing the decision of the expert, one

party considers that the expert got it wrong. To apply to the Court in anticipation of the decision is likely to prove wasteful of time and costs.

The salient features of the various processes, summarised as expert determination, are as follows:

- The expert is not bound by the Arbitration Acts, in particular there is no statutory right of appeal or determination of a preliminary point of law.
- The court has no statutory powers to make interlocutory orders in aid of an expert.
- An expert's determination is not enforceable as a judgement in the same way as an arbitrator's award.
- The expert makes a decision on own expertise and investigations, and is not bound to receive evidence or submissions from the parties, but is not in control of procedure, and must comply with the terms of the underlying contract from which authority is derived.
- The expert is not bound to act judicially, but merely to avoid fraud or collusion and may also be liable to the parties for negligence.
- The expert's fees are only recoverable as a debt in the absence of an express term conferring on the expert the ability to award costs against a party.

Expert determination is a private means of commercial dispute resolution. The parties to a contract jointly instruct a third party to decide a dispute. The essentials of the role of the expert are:

- No requirement to act judicially but may carry out investigation.
- No requirement to receive submissions or evidence.
- No legislation to cover process.
- No statutory right of appeal or review.
- No statutory right of registration of and enforcement of decision.

However, the expert must act in accordance with the terms of his appointment (which may have implied terms or incorporate parts of the contract) if his award is to be valid.

There is nothing new about expert determination. Kendall[13] reports that it has been a feature of English commercial and legal practice for at least 250 years. It is:

- A process by which parties instruct a third party to decide a particular issue.

- A creature of contract.
- The third party is selected for particular expertise.
- Courts are reluctant to intervene, and have restricted to jurisdiction (*Jones* v. *Sherwood* 1992), or where the expert has answered the wrong question.

The courts have recently considered the procedure of *expert determination* and have reiterated and confirmed earlier judgements.[14]

Contracted mediation

Contracted mediation is project-based mediation. A panel (provided for in the contract) is set up at the start of a project and actively seeks to facilitate the avoidance and resolution of any contract difference throughout the project, before it can escalate into dispute and before parties have started incurring any costs. Contracted mediation attempts to fuse team building, dispute avoidance and dispute resolution in one procedure. The impartial contracted mediation panel, perhaps consisting of one lawyer and one commercial expert who are both trained mediators, is appointed at the outset of the project. The panel attends site meetings and conducts workshops. The panel members therefore gain a working knowledge of the project and all actors and stakeholders involved in and working on the project. This knowledge they assimilate allows the panel to resolve contractual differences before they escalate, and provides for the confidential, mediated resolution of disputes. The panel have access to the contract documents and the parties involved.

Contracted mediation has been used in Jersey Airport,[15] where a civil engineering project was the first contract undertaken with a contracted mediation panel in place from the outset to help prevent, mitigate and resolve disputes.

Dispute review boards, dispute review panel, dispute adjudication board and dispute adjudication panel

This is a particularly difficult technique because there are many techniques and the names are very similar. Beware! Dispute review boards led the way and a description is given below, but the variants are many; for example the International Federation of Consulting Engineers (FIDIC) uses *dispute adjudication boards*, the 2012 London Olympics set up an *independent dispute avoidance panel* AND, for any dispute not avoided, a *dispute adjudication panel*. If confronted with this kind of technique it is sensible to consider the power of the board or panel as agreed in the contract. Is it binding or non-binding? Is it binding until completion then something else, i.e. is it temporary binding, or is binding forever?

Dispute review boards develop in the USA as a process where an independent board evaluates disputes as they arise during the project being constructed or manufactured, and makes settlement recommendations to the parties. The board is normally constituted at commencement of the project and a typical scenario is where one board member is selected by each party and they agree on an independent third – a referee or umpire. Failure by the parties to agree on the umpire normally requires that the two board members then select the third member. The board periodically visit the site and receive project information to ensure familiarity with the project and the parties. The board meets regularly to discuss problems or disputes, hears presentations from parties and suggests solutions.

Dispute review boards derive their authority from the underlying contract and may be conferred with the power to produce advisory, interim binding or finally binding decisions. The principal difference between a DRB and an executive tribunal will be the constitution of the board/tribunal – the former will usually be independent of the parties, the latter will include representative of each party.

The concept of DRB is usefully explained at the dispute resolution board foundation.[16] There are nine essential elements necessary for a DRB to be successful. If any of these elements are missing, success is jeopardised. These elements are:

1 All three members of the DRB are neutral and subject to the approval of both parties.

2 All members sign a three-party agreement obligating them to serve both parties equally and fairly.

3 The fees and expenses of the DRB members are shared equally by the parties.

4 The DRB is organised when work begins, before there are any disputes.

5 The DRB keeps abreast of job developments by means of relevant documentation and regular site visits.

6 Either party can refer a dispute to the DRB.

7 An informal but comprehensive hearing is convened promptly.

8 The written recommendations of the DRB are not binding on either party but are admissible as evidence, to the extent permitted by law, in case of later arbitration or litigation.

9 The members are absolved from any personal or professional liability arising from their DRB activities.

Dispute resolution adviser (DRA)

The concept of DRA was developed by many commentators independently; a useful description of the many and varied techniques can be found in Wall.[17] Colin Wall is a dispute specialist in Hong Kong who has used the technique to run complicated and difficult construction projects with many problems and complete with few major disputes. The system of DRA run in Hong Kong will be described after the general themes common to the DRA concept have been introduced.

The concept

Colin Wall suggests that the idea of a dispute resolution adviser (DRA) came from Clifford Evans, who, in 1986, suggested the use of an independent intervener. This independent intervener would be paid for equally by the employer and the contractor to settle disputes as they emerged, rather than wait until the end of the contract. Any decision would be binding until the end of the project and then either party could commence arbitration proceedings. Wall converted the independent intervener into the DRA who does not make interim binding decisions but advises on the means by which a settlement could be achieved. Power to settle rests with the parties.

The DRA's first action is to deal with disagreements at site level, so that these can be addressed before they develop into full blown disputes. This avoids the breakdown in working relationships that often takes place and then effects the rest of the project, and it allows the issues to be dealt with while they are fresh in the party's mind. Neither the parties or the DRA are limited to legal decisions and commercial settlements could encompass wider solutions which are mutually beneficial to all.

The DRA system

The system developed by Wall in Hong Kong became the DRA system and was used on Hong Kong Government Projects, notably the Queen Mary Hospital. The DRA system has distinct stages, as shown in Figure 8.1 and is described as follows:

> DRA system draws upon the independent intervener concept as modified by the dispute adviser but provides a more flexible approach. It embodies the dispute prevention techniques of the dispute review board and project arbitration, it uses a partnering techniques to re-orient the parties' thinking and encourages negotiation by using a tiered dispute resolution process. It is based on giving parties maximum control by the use of mediation techniques but also includes prompt short form arbitration which encourages voluntary settlement and if necessary provides a final and binding resolution to a dispute.

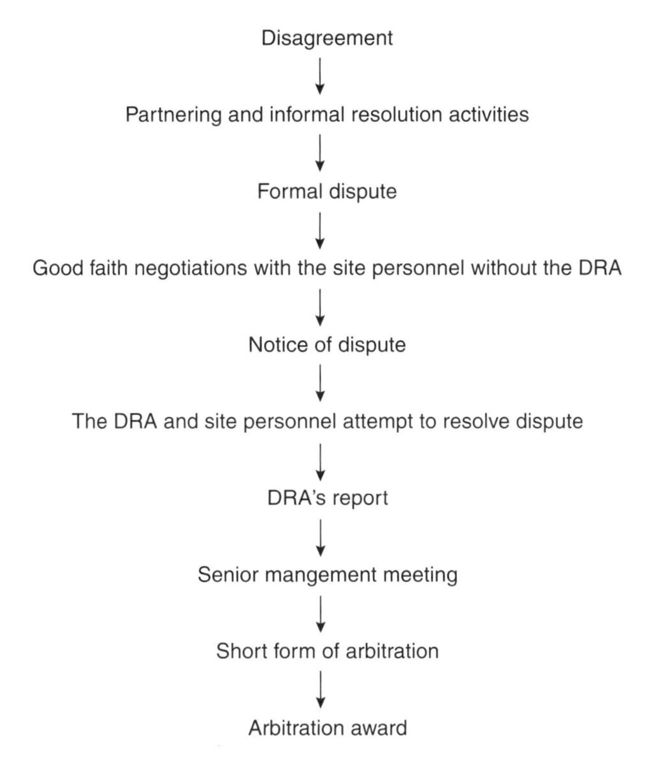

Figure 8.1 An outline of the DRA system

At the commencement of the project the DRA undertakes partnering styled activities focused on building relationships with the parties and that the same time encouraging teamwork. During the period of the project the DRA visit the site on a regular basis in order to maintain a level of familiarity with the project and the parties. At this stage, the DRA also assists on the settlement of any disagreements, which may have arisen between visits. During any formal disputes that occur the DRA attempts to facilitate their settlement using a variety of techniques from mediation to expert determination in particular areas. The DRA attempts to maintain a purely facilitative role in order to sustain the impartial and neutral position, which has developed; therefore any evaluation is carried out by another neutral third party and not the DRA.

If disputes proceed beyond settlement at site level, the DRA produces a report outlining the nature of the dispute and the parties' cases. This report may contain a non-binding recommendation or an evaluation of the dispute. Parties are given opportunities to check the accuracy of the report and comment providing an important chance for them to review their position before the report is passed to senior management. Senior management will then be able to obtain a clear picture of the dispute and hopefully bring an unemotional perspective to the problem. Again, at this level the DRA may continue to facilitate the resolution of the dispute by assisting the senior managers.

If the matter remains unresolved 14 days after the DRA report then a short form of arbitration may be employed. This takes place within 28 days of the end of the senior management efforts, and the arbitrator is selected by the parties, perhaps assisted by the DRA. The rules of the short-form arbitration include the following key elements:

- One issue or a limited number of issues, conducted in one day per issue.
- Each party is given the opportunity to present.
- Each party is given equal time to be present.
- The arbitrator produces an award in a very short period which is final and binding.
- Dispute concerning time or money are resolved using final offer arbitration where the arbitrator must select one or the other picture.

The Queen Mary Hospital project raised numerous problems yet there were no disputes that proceeded to short form arbitration, producing a large saving in time and money.

9 Cooperation and collaboration

Pink ribbon

Should a distinction be made between cooperation and collaboration? It is clear that the 'fashion' of the time is both. No distinction is made here, and dictionary definitions are very similar, although usage often starts to indicate that collaboration introduces more sinister implications. That is, cooperation indicates healthy teamwork activities, while collaboration begins along the road of collusion and conspiracy that ends with monopoly. This chapter considers how the trend towards cooperation (for example via partnering or alliancing) came about, how it affects conflict and dispute, and why competitive organisations would cooperate and the 'third' way of combining cooperation and competition. This third way has been described as co-opetition, where the best features of both competition and cooperation are combined.

This chapter introduces concepts of game theory, which are taken further in Chapter 10.

In a competitive environment, when would cooperation break out? And why? It seems counterintuitive – in dynamic capitalism competitors cooperate! It goes against all reasoning or logic. Perhaps the nature of man as the only philanthropic and altruistic animal is more right than we thought.

Examples are drawn from a very famous book, *The Evolution of Cooperation*, in which the argument is that the evolution of cooperation requires that firms and organisations will cooperate if there is a sufficiently large chance that they will meet again in order that they have a stake in their future interaction. Given these circumstances, cooperation evolves in three stages:

1 Cooperation will be initiated if firms have the opportunity for future interaction. Cooperation will NOT be initiated by scattered firms who have little chance of future interaction.

2 Cooperation will thrive where many other strategies are being tried.

3 Cooperation, once established, will protect itself from competitive strategies.

This cooperation theory has many organisational and political applications, not least for the purposes of this book, the cooperation of commercial organisations thereby reducing conflict and disputes.

Examples are given of cooperation in the most extraordinary circumstance: cases of 'live and let live' in the First World War, where opposing troops simply did not shoot each other, providing that the restraint was reciprocated. In Northern Ireland after years (centuries) of intractable conflict cooperation broke out. A brief case study of a major alliance in the airline sector is given.

Introduction

Cooperation and collaboration are currently popular methods to manage conflict and avoid disputes. This trend comes from many things, including: economic theory, contract (legal) theory and game theory.

Economic theory treats people as rational beings, but clearly human beings do not always act rationally. To cope with this irrationality economist developed theories of bounded rationality – the idea that in decision making, rationality of individuals is limited by the information they have, their minds and the finite amount of time they have to make decisions. 'Traditional' economic theory also dictates that markets are efficient and, as a result, firms will always contract out the work rather than make. This is often described as the make or buy decision. Transaction cost analysis changed this thinking and the plethora of cost associated with 'buying', when factored in, change the dynamics of make or buy.

Contract theory, where law and economics come together, studies how parties and stakeholders construct contractual arrangements. This is generally subject to what is described as asymmetric information. That is where one party has more or better information than the other. This creates an imbalance of power in transactions which can sometimes cause the transactions to go wrong. If you seek a theoretical background to this book, there you have it: the parties to a commercial contract often have information asymmetry which causes conflict. A logical development from the make or buy decision is towards firms combining to cooperate.

Cooperation and collaboration

Does a distinction need to be made? Again, the problem that has dogged this book rears its head – the problem of definition. The dictionary definitions of cooperation and collaboration are very similar, but there are usage hints which begin to indicate that collaboration introduces more sinister implications. Cooperation indicates healthy teamwork activities, while collaboration begins along the road of collusion and conspiracy that ends with monopoly. No distinction is made in this book, but there is debate – for a summary try reading Yeung *et al.* (2007).[1] The term collaboration is not used here.

Dynamic capitalism dictates competition and that firms will not cooperate; they have no incentive, in fact the incentive is to be selfish. But cooperate they do, and examples include partnering, alliancing and relational contracting all forms of a generic collaborative working or contracting.

Partnering, alliancing and relational contracting

As with lots of developments in contracts and procurement, it is difficult to find clear definitions and therefore to distinguish between these three areas which are sometimes collected under the heading of relational contracting. So with that in mind, the following definitions are offered which might both be seen as developing concepts of collaborative working.

Partnering – this is very simple, it is just people working together – a voluntary system of handling normal, everyday commercial problems in a mutually agreeable manner before they turn into major issues that create disputes. Partnering is normally either: strategic (long term) or project (short term). Many feel that strategic partnering is inappropriate for public sector or government because of accountability issues.

Alliancing – a relationship entered into by two or more parties in which their interests are aligned on a legally enforceable basis of shared gain/pain and 'no blame' to deliver outstanding project performance.

Relationship contracting – a process to establish and manage the relationships between parties that aims to remove barriers, encourage maximum contribution and allow the parties to achieve success.

So, these definitions have concepts of cooperation at their heart. You can read many more definitions and theory in the literature. Relationship contracting is based on achieving successful project outcomes, including:

- Completion within budget.
- Completion on time.
- Strong people relationships between the parties resulting from mutual trust and cooperation, open and honest communication.

- Optimum project life-cycle cost.
- Achieving optimum standards during the project lifetime for:
  - safety
  - quality
  - industrial relations
  - environment
  - community relations.

Relational contracting establishes a business relationship which is designed to deliver optimum commercial benefits to all the parties. It is founded on the principle that that cooperation delivers a mutual benefit to the parties and they are incentivised to deliver the project at the optimum cost, on time and to the quality specified. To achieve this, the relationship between the client and the contractor cannot be taken for granted, even if they have worked together before and established a close business relationship it is still crucial that they build the relationship for each specific project. In order to do this, the relationship must be founded on strong mutually held core values and guiding principles. These are summarised above in Table 9.1.

The success of relational contracting depends on the willingness of both parties to commit to the change at an individual and project level.

The fundamentals of relationship contracting[2]

A list of the fundamentals of relationship contracting shows some familiar issues:

1. Alignment of goals.
2. Risk allocation.
3. Clearly defined project scope.
4. Form of contract.
5. Integrated project team.
6. Gainshare/painshare.
7. Open honest communications/behaviour/change of attitude.
8. Public sector issues.
9. Facilitators.
10. Legal advisers.
11. Third party advisers.

Table 9.1 Core values and guiding principles for relationships

Core values	*Guiding principles*
Commitment	Total commitment to achieving the project goals which is actively promoted by senior management of all parties
Trust	To work together in a spirit of good faith, openness cooperation and not to seek to apportion blame
Respect	The interests of the project take priority over the interests of the parties
Innovation	To couple innovative or breakthrough thinking with intelligent risk taking to achieve exceptionally good project outcomes.
Fairness	To integrate staff from all parties on the basis of fairness and the best qualified for the job
Enthusiasm	To engender enthusiasm for professional duties and the project's social activities

Conflict and dispute in relational contracting

Relational contracts have not eradicated contract dispute, and this is clear from the contracts which include the usual panoply of dispute resolution provisions. See for example the Joint Contracts Tribunal (JCT), *Constructing Excellence Contract*, one of the range of JCT contracts which includes the standard dispute resolution provisions. Further evidence is available from the literature where the usual papers, academic and practitioner, debate the range of dispute resolution provisions, for example see Yates and Duran.[3]

The problems and benefits of cooperation

The problems of cooperation might be seen as the absence or lack of benefits that go with competition. These include:

1 Lower prices for consumers.
2 Discipline on supplier firms in the market (which assists them to keep or drive their costs down).
3 Improvements in technology – with positive effects on production methods and costs.
4 Consumer choice.
5 Supplier invention and innovation.
6 Improvements to the quality of service for consumers.

7 Increased (and better) information for consumers allowing people to make more informed choices.

On the other hand, the benefits of cooperation include:

1 Greater efficiency.
2 Economies in tendering.
3 Increased (and better) information for suppliers allowing them to make more informed tender choices.
4 Economies of and in scale.
5 Access to cooperators expertise.

A third way: co-opetition

Has it got to be competition OR co-operation? The holy grail is to attain the benefits of both without losing too much, without gaining too many of the problems. Game theorists describe this as co-opetition.[4] Co-opetition occurs when companies work together. They may share common parts of their business where they do not believe they have competitive advantage and where they can share common costs. Many examples of companies employing co-opetition are given and this is a fast-developing area.

There are further examples – public sector procurement groups, such as those developed in Northern Ireland recently. In addition, the development of public sector mergers and cooperation whereby, for example, groups of housing associations or charities work together under formal agreements to develop ways of combining their resources or buying power while retaining their independence in terms of the services that they provide. Those bigger group structures are notoriously difficult to develop because of the perceived competition between organisation and because no small independent public sector organisation wants to lose their 'ethos' in a much larger group.

When will cooperation break out

This section draws heavily on a famous book, which I implore you to read, Axelrod's *The Evolution of Cooperation*.[5] The question is repeated directly from that book:

> Under what conditions will cooperation emerge ... this question has intrigued people for a long time. And for good reason. We all know that people are not angels, and that they tend to look after themselves and their own first. Yet we all know that cooperation does occur and our civilisation is based upon it. But in situations where each individual has an incentive to be selfish, how can cooperation ever develop?

The answers to these questions have formed our philosophy, our politics and our systems of government. Great attention is given to the absence or presence of a central controlling authority, and arguments about the correct and proper scope of government have centered on whether cooperation could be expected without an authority to police the situation. *The Evolution of Cooperation* found that cooperation, based on reciprocity, can emerge if firms have a chance to meet again. So that they have a stake in their future interaction. However, cooperation will NOT be initiated by scattered firms who have little chance of future interaction. Once initiated, cooperation will thrive where many other strategies are being tried, and, once established, cooperation will protect itself from competitive strategies.

So, cooperation will break out if there are sufficient firms in the marketplace so that firms have a chance of affecting any future interaction. Once it is up and running, cooperation will thrive where many other strategies are being tried, and once running cooperation established will protect itself from competitive strategies.

Examples of extraordinary cooperation

1 *First World War cases of 'live and let live'*

During the First World War there are famous examples of the troops holding ceasefires at Christmas, culminating in games of football between the opposing troops in no man's land. What is less widely known is the extraordinary cooperation over long periods where opposing troops simply did not shoot each other providing the restraint was reciprocated.[6]

Cooperation broke out because the opposing troops were so close they could see each other and, in a way, knew each other, and restraint was initiated and reciprocated. Axelrod paints it as an example of the iterated prisoner's dilemma (see the Chapter 10).

Considerable etiquette developed, Axelrod describes how complex rules were laid down and elaborate apologies delivered. The following lists some of the quotations, full details can be found in *The Evolution of Cooperation*.

Live and let live

There are examples of extraordinary cooperation. During the First World War the Western Front (a five-hundred-mile line in France and Belgium) was the scene of bloody battles for a few yards of territory. Many lives were lost for a few yards of advance, and then those gains were surrendered. However, between those advances and retreats, and sometimes even during them, at many places along the Western Front the enemy soldiers often exercised considerable restraint – going as far as simply not shooting at each other. This became known as live and let live.

A British staff officer on a tour of the trenches remarked that he was

> astonished to observe German soldiers walking about within rifle range behind their own line. Our men appeared to take no notice. I privately made up my mind to do away with that sort of thing when we took over; such things should not be allowed. These people evidently did not know there was a war on. Both sides apparently believed in the policy of 'live and let live'.
>
> (Dugdale 1932)[7]

This is no isolated example. The live and let live system was widespread in trench warfare and it flourished. The officers tried to stop it but it persisted. How can cooperation emerge despite great antagonism between the players? Perhaps the conscripts were not persuaded by the view of the officers and used tactics within their power to cooperate. The average infantryman confirmed the view that mindless slaughter was not in their interests and should be avoided by cooperation.

> The real reason for the quietness of some sections of the line was that neither side had any intention of advancing in that particular district ... If the British shelled the Germans, the Germans replied, and the damage was equal: if the Germans bombed an advanced piece of trench and killed five Englishmen, an answering fusillade killed five Germans.
>
> (Belton Cobb 1916)[8]

How did this cooperation get started and how did it establish and propagate itself? Early in the war it seems there were times in the day when cooperation to avoid fighting would benefit all sides. Meal times were an obvious example:

> The quartermaster used to bring the rations up ... each night after dark; they were laid out and parties used to come from the front line to fetch them. I suppose the enemy were occupied in the same way; so things were quiet at that hour for a couple of nights, and the ration parties became careless because of it, and laughed and talked on their way back to their companies.
>
> (King 1938)[9]

These meal time cooperations flourished and led to the famous Christmas fraternisations of 1916. The following months brought direct truces which were arranged by simple shouts or by signals. 'In one section the hour of 8 to 9 a.m. was regarded as consecrated to "private business", and certain places indicated by a flag were regarded as out of bounds by the snipers on both sides' (Morgan 1916).[10]

> It would be child's play to shell the road behind the enemy's trenches, crowded as it must be with ration wagons and water carts, into a blood-stained wilderness ... but on the whole there is silence. After all, if you prevent your enemy from drawing his rations, his remedy is simple: he will prevent you from drawing yours.
>
> (Hay 1916)[11]

2 *Northern Ireland*

The history of Northern Ireland has occupied much better texts than this, and any attempt at summarising would be bound for failure and to cause offence. Suffice to say, after years (centuries) of intractable conflict and suffering in Northern Ireland, cooperation broke out. The Northern Ireland example is held up as how optimism must persist – if they can form an agreement there and cooperate, then agreement and cooperation is possible anywhere. It is cooperation and agreement of optimism.

Alliancing case study: Star Alliance a global airline alliance

The Star Alliance network was established in 1997 as the first truly global airline alliance to offer worldwide reach, recognition and seamless service to the international traveller. Its acceptance by the market has been recognised by numerous awards, including the Air Transport World Market Leadership Award, Best Airline Alliance by both *Business Traveller Magazine* and Skytrax. The member airlines are Adria Airways, Aegean Airlines, Air Canada, Air China, Air New Zealand, ANA, Asiana Airlines, Austrian, Blue1, bmi, Brussels Airlines, Continental Airlines, Croatia Airlines, EGYPTAIR, LOT Polish Airlines, Lufthansa, Scandinavian Airlines, Singapore Airlines, South African Airways, Spanair, SWISS, TAM Airlines, TAP Portugal, Turkish Airlines, THAI, United and US Airways. Air India, Avianca-TACA, Copa Airlines and Ethiopian Airlines have been announced as future members. Overall, the Star Alliance network offers 21,000 daily flights to 1,160 airports in 181 countries.

Star Alliance has 70 staff, managing the coordination of the alliance of 27 airlines with a combined turnover of $157bn. A fundamental tenet of the Alliance is innovation in their service provision, which began with the development of a Technology Innovation Advisory Council in 2008; this featured world-leading technology vendors and academic input from Massachusetts Institute of Technology. The technology-led approach brought success and consequently in 2010, the Advisory Council evolved into a corporate innovation group, with a more bottom-up approach of encouraging and promoting staff involvement in innovation, creating *real solutions to actual problems*. One of the first activities this group undertook was to form an *innovation community* of volunteers from the member airlines around a web-based system, currently with 1,100 members of varying levels of participation. Primarily, this has supported idea generation and concept development, although already some ideas from this community have moved into product development phase.

10 Game theory

Pink ribbon

The previous chapter used a technique developed from game theory: co-opetition. This chapter describes more of the history of game theory and discusses the ancient and modern views on strategy. Recent thinking links game theory with strategic thinking. Game theory is the art of anticipating an opponent's next move, and understanding that a competitor is attempting the same thing.

One of the most quoted texts on strategy is an ancient Chinese military treatise that is attributed to Sun Tzu.[1] Sun Tzu is often quoted in business and management and reflects the thinking that says commerce is similar to war. A quotation from Gore Vidal is often given:

> It is not enough to succeed. Others must fail.[2]

This presented in game theory terms as win–lose.

Current thinking links game theory and strategy.[3] The game theory applications to strategy (particularly co-opetition) reflect the current thinking that commerce is similar to life: we have to get along, there is no alternative. The quotation to illuminate this is from a financier Bernard Baruch:

> You don't have to blow out the other fellows light to let your own shine.[4]

This presented in game theory terms as win–win.

Game theory might then be thought as being at the heart of conflict management and dispute resolution, for if the only option was win–lose, then the only option would be dominance. Throughout this book the option of integration is pursued. Mary Parker Follett started this in the early part of the twentieth century with ideas of integration in conflict, where each side refocuses their efforts so that the problem is solved and neither side loses. Negotiation theory developed this integrative approach into a principled approach where focus is on the parties' interests and not their positions. Facilitative mediators developed skills whereby their presence encouraged this.

Game theory has developed in many areas to foster new solutions: in economics, business, law, science and engineering. What, then, can it offer conflict management and dispute resolution? Everything. Predicting the future is difficult (see Chapter 2) but the success of game theory in other fields must mean new developments in conflict management and dispute resolution. Examples are given here of game theory in negotiation (tacit bargaining) and game theory in law (information escrow).

History

Although there were discussions before John von Neumann in the early part of the twentieth century, von Neumann is usually credited as the inventor of game theory via his work culminating in a book from 1944, *Theory of Games and Economic Behavior*, by von Neumann and Morgenstern.[5]

This book showed, among other things, that social interaction and events can be described, and explained, by models taken from games of strategy. Two types of game are normally described: zero-sum and non-zero-sum.

In zero-sum games, an individual does better at another's expense. A zero-sum game is a mathematical representation of a situation in which a participant's gain or loss is exactly balanced by the losses or gains of the other participant(s). If the total gains of the participants are added up, then the total losses are subtracted, they will come to zero. The often quoted example is one from principled negotiation – cutting a pie is a zero-sum game because taking a larger slice reduces the amount of pie available for others. Zero-sum games are most often solved with the minimax theorem from von Neuman's theory, or with Nash equilibrium from another famous game theorist John Nash.

In non-zero-sum games the situation changes and becomes where one gain (or loss) does not necessarily result in another's loss (or gain). Where the winnings and losses of all players do not add up to zero and everyone can gain. This is the derivation of the truly awful phrase 'win–win situation'. The principled negotiation talk is of expanding the pie so that everyone can get more. A famous example of a non-zero-sum game is the prisoner's dilemma.

The early work characterised by von Neumann contains the method for finding mutually consistent solutions for two-person zero-sum games. This early game theory concentrated on cooperative games and the optimal strategies for groups of individuals with the presumption that they can enforce agreements.

In the 1950s, game theory was much used in strategy issues and its applications were focused on the cold war. About this time the first discussion of the prisoner's dilemma appeared, and an experiment was undertaken on this game at the RAND Corporation. The RAND Corporation is a non-profit organisation best described by its mission:

> The RAND Corporation is a non-profit institution that helps improve policy and decision making through research and analysis.
>
> RAND focuses on the issues that matter most such as health, education, national security, international affairs, law and business, the environment, and more. With a research staff consisting of some of the world's preeminent minds, RAND has been expanding the boundaries of human knowledge for more than 60 years.
>
> (www.rand.org)

As a non-partisan organisation, RAND attracts wide respect for operating independent of political and commercial pressures. Of course, much more can be found at the RAND website (www.rand.org).

Around the same time, John Nash developed theories around non-competitive games as well as von Neumann's work on competitive games. The most famous was known as Nash equilibrium, and Nash's life was summarised in biography and a popular film *A Beautiful Mind*.[6] This had much wider application to a wider variety of games than the work of von Neumann.

Throughout the 1960s, game theory continued its development and the first of many Nobel Prizes was awarded. Nobel Prizes are a useful performance indicator to the importance of an academic discipline, and game theory has been a fertile ground for Nobel Laureates.

In the 1970s, game theory started to see widespread application in other disciplines notably at first in biology. Game theory is now used extensively in many disciplines. So much, in fact, it is all things to all men – it is human behaviour, mathematics, economics, philosophy, computer science, logic. This chapter considers game theory in terms of the behaviour of firms under conflict and dispute.

There has been considerable recent interest in the use of game theory to explain firms' behaviour, and one measure is the Nobel Prizes awarded for work in the area. While there are many examples of game theory applications, the most famous game theory example is the prisoner's dilemma.

The prisoner's dilemma

The game gets its name from the following hypothetical situation: imagine two criminals arrested under the suspicion of having committed a crime together. However, the police do not have sufficient proof in order to have them convicted. The two prisoners are isolated from each other, and the police visit each of them and offer a deal: the one who offers evidence against the other one will be freed. If neither of them accepts the offer, they are in fact cooperating against the police, and both of them will get only a small punishment because of lack of proof – they both gain. However, if one of them betrays the other one, by confessing to the police, the defector will gain more, since he is freed, and the one who remains silent will receive the full punishment, since he did not help the police, and there is sufficient proof. If both betray, both will be punished, but less severely than if they had refused to talk. The dilemma resides in the fact that each prisoner has a choice between only two options, but cannot make a good decision without knowing what the other one will do.

Ancient and modern views on strategy

A common definition of tactics and strategy points towards the military influence in their development. *Tactics win the battle: strategy wins the war.* Sun Tzu was an ancient military figure in China from the eighth century BC. Sun Tzu was a general, a strategist and philosopher and is widely believed to have authored *The Art of War*, perhaps the most famous military treatise of all. Other developing fields with an interest in strategy, for example politics and management, borrowed from military strategy and tactics. Niccolò Machiavelli wrote a famous political treatise *The Prince*. It is intriguing how both books have remained so influential and Machiavelli's name has passed into our language as a description of a person with a cunning or duplicitous disposition – a machiavellian. *The Art of War* is quoted and used frequently, and here is Sun Tzu on strategy and tactics: 'Strategy without tactics is the slowest route to victory. Tactics without strategy is the noise before defeat'.

Management thinking picked up these ideas from Sun Tzu and there exists a view that business and commerce is like war, therefore strategy in commerce is redolent with Sun Tzu's thinking. Game theory argues against this, as Chapter 9 indicated. While military strategy might be considered win–lose, game theory offers a solution that allows both sides to win: win–win.

The Art of War

Of course, the danger in any document more than two thousand years old and written in another language is that nuances are lost over time and in translation. The headings given here for chapters are from three separate translations.[7]

Chapter 1: Laying Plans/The Calculations/Detail Assessment and Planning
This chapter explores the fundamental factors and elements that determine the outcomes of military engagements. Sun Tzu stresses that war is a very grave matter for the state.

Chapter 2: Waging War/The Challenge/Waging War
This chapter explains the economics of warfare, success requires winning decisive engagements quickly. In a move centuries before others have become concerned with cost, Sun Tzu advises that successful military campaigns are aware of the cost of competition and conflict.

Chapter 3: Attack by Stratagem/The Plan of Attack/Strategic Attack
Here, Sun Tzu defines the source of strength as unity not size.

Chapter 4: Tactical Dispositions/Positioning/Disposition of the Army
Sun Tzu explains the importance of defending existing positions until capable of advancing from those positions in safety.

Chapter 5: Energy/Directing/Forces military manoeuvres
Sun Tzu advocates the use of creativity and timing.

Chapter 6: Weak points and strong/Illusion and reality/Weaknesses and strengths
Sun Tzu explains how opportunities arise caused by the relative weakness of the enemy in a given area.

Chapter 7: Manoeuvring/Engaging the force/Military manoeuvres
Sun Tzu explains the dangers of direct conflict and how to win confrontations when they are forced.

Chapter 8: Variation in tactics/The nine variations/Variations and adaptability
This chapter describes the need for flexibility in responses.

Chapter 9: The army on the March/Moving the force/Movement and development of troops
Sun Tzu describes how to respond to differing situations. This chapter pays great attention to evaluating the intentions of others.

Chapter 10: Terrain/Situational positioning/Terrain
Sun Tzu looks at general areas of resistance and the six types of ground positions that arise from them. Here is one of Sun Tzu's most famous sayings:

> It is a military axiom not to advance uphill against the enemy.

Chapter 11: The nine situations/The nine situations /The nine battlegrounds
Sun Tzu describes the nine common situations in a campaign, from scattering to deadly, and the specific focus required to successfully navigate them.

Chapter 12: The attack by fire/Fiery attack/Attacking with fire
Sun Tzu explains the general use of weapons and the specific use of the environment as a weapon.

Chapter 13: The use of spies/The use of intelligence/Intelligence and espionage

Sun Tzu focuses on the importance of developing good information sources and managing information.

Sun Tzu is popular and has been rediscovered many times. Business books apply Sun Tzu's lessons to many things, for example office politics and corporate strategy.[8] It is reported that Japanese companies make the book required reading for their key executives.[9]

Other than arguing against the employment of warlike (military) imagery in commerce and business, game theory has many techniques, ideas and benefits. Just two are given here: tacit bargaining and information escrow.

Tacit bargaining

Tacit bargaining[10] was first formally described by Thomas Schelling, another Nobel Laureate, to cover those examples in negotiation that 'just' happen. Situations where people have intuitive information, where bargaining takes place in which communication is incomplete or impossible. Schelling gives the following examples:

- When asked to pick any number at random, four in ten people picked the number one.
- When asked to predict a coin toss the overwhelming majority call heads.
- When asked to pick an amount of money almost all people chose a figure divisible by ten.
- When people were told that they had to meet someone else – but had to guess the time – almost all chose noon.

Conversely, many people have an incorrect view on combined probabilities. The tale is told of a bidder for the UK National Lottery. As a unique selling point (USP) the bidder suggested that their bid would make every player who guessed six numbers in the lottery a millionaire. It was pointed out that each week 20,000 chose to play the numbers 1, 2, 3, 4, 5 and 6. And that if that sequence came up the company would not be able to honour the prize. *But 1, 2, 3, 4, 5 and 6 will never come up*. It was then pointed out that this combination was just as likely as any other and that the financers could not back the proposal.

People concede to convention without even knowing it, tacitly and noticeably when faced with unfamiliar customs. It is natural to want to abide by notions of fairness and precedence. Don't rock the boat. This is further examined in psychology as groupthink, which occurs within groups of people. Group members try to minimise conflict and reach a consensus decision without critical evaluation of alternative ideas or viewpoints.

The importance for negotiation is that old question: Who should make the first offer? Counter intuitively, it is generally an advantage to make the first move in a negotiation, even though most people are reticent to do this. Making the first move gives a party the opportunity to frame the negotiation and establish precedence.

Further insight gained from research into tacit bargaining is that an outsider can often be more effective, sales teams will often include starters and closers. The starter begins the sales pitch and passes the customer to the closer. The customer sees the closer as an outsider and is more agreeable to the closing.

Information escrow

Information escrow was proposed by game theorists at Harvard[11] and has immediate applicability in negotiation which has been picked up by mediators particularly those offering settlement mediation techniques (see Chapter 5).

Escrow comes from an old French word (meaning scrap of paper) and is best known in American legal speak. It generally refers to a trusted third party who holds something on trust. The usual example is funds held in a separate account.

A formal definition might be an arrangement where an independent trusted third party receives and disburses money and/or documents for two or more transacting parties, with the timing of such disbursement by the third party dependent on the performance by the parties of agreed-upon contractual provisions.

In an information escrow, each side makes an offer to a neutral third party. As an example, the parties to a dispute appoint a third party neutral to receive their offers. The claiming party offers to accept £3m to settle the matter privately and in confidence to the third party. The other party, privately to the third party, offers to pay up to £5m to settle the matter. Since the offers cross, a deal may be done at £3m and settlement reached. If the offers do not cross then neither party learns what the other's offer was. In addition, the neutral third party may utilise facilitative techniques to attempt to bring the parties together in other ways. Some worry about the ethical problems that face the third party neutrals since they alone know the offers, and, in this example, why settle at £3m not £4.95m?

Conclusion

Game theory might then be thought of as being at the heart of conflict management and dispute resolution, for if the only option was win–lose then the only option would be dominance. Throughout this book the option of integration is pursued. This started in Chapter 2 where Mary Parker Follett's development to functional conflict started in the early part of the twentieth century with ideas of integration in conflict, where each side refocuses their efforts so that the problem is solved and neither side loses. Chapter 4 describes how negotiation theory developed this integrative approach into a principled approach in *Getting to Yes*, where focus is on the parties' interests and not their positions. Chapter 5 continues this; facilitative mediators developed skills whereby their presence encouraged integration. Chapter 9 describes how the game theory application of co-opetition suggests that cooperation and collaboration really are the only hope. Game theory has developed in many areas to foster new solutions: in economics, in business, in law, in science and engineering. What, then, can it offer conflict management and dispute resolution? Everything. Predicting the future is difficult (see Chapter 2) but the success of game theory in other fields must mean new developments in conflict management and dispute resolution. This is a developing area – put aside your cynicism, but not your scepticism, and embrace the developments.

11 The stages of dispute resolution

Pink ribbon

It is useful to talk about dispute resolution in stages. Using the classification system developed earlier it might be the stages of conflict management and dispute resolution. There then become four stages:

1 Conflict management (or dispute avoidance).
2 Negotiation as dispute resolution.
3 Non-binding (or consensual) dispute resolution.
4 Binding (or adjudicative) dispute resolution.

Alternatively, some authors talk of three core techniques, which may be employed in the resolution of disputes:

1 Negotiation.
2 Third party intervention.
3 Adjudicative process.

Negotiation refers to the problem-solving efforts of the parties, third party intervention does not lead to a binding decision being imposed on the parties, and finally the adjudicative process, the ultimate outcome, is an imposed binding decision.

Others talk of the *three pillars* of dispute resolution. The discrete techniques can be introduced under one of the three pillars, depending upon the main characteristics of the particular technique.

The UK Government's approach is to talk of principal stages and to include techniques within the stages (which are not named). This allows the major techniques to be listed and reminds that negotiation can continue through all three stages.

Stage 1: Negotiation.

Stage 2: Mediation, conciliation, neutral evaluation, construction adjudication.

Stage 3: Arbitration, expert determination, litigation.

Some have taken this approach into contracts and include tiered dispute resolution clauses. Here, when disputes arise, they are dealt with in stages, starting with negotiation and if this cannot resolve the matter, a third party intervention technique is employed. Finally, an adjudicative process is used. Project-based industries have a need to keep the project 'on-track' if a dispute arises and employ techniques to provide for Stage 2 techniques, which provide an adjudicative decision that is binding until the project is complete.

Introduction

Common sense says that dispute resolution needs to be dealt with in stages. In Chapter 2 we considered the UK Government's approach (see Figure 2.2). Here, the major techniques are listed in unnamed stages: Stage 1, Stage 2 and Stage 3.

Stage 1: negotiation

In Stage 1 the technique is negotiation, but it is clearly shown that negotiation can continue throughout all dispute resolution stages. This is a useful reminder – negotiation is clearly the first technique which should be tried – it is the cheapest and it is something everyone can do (see Chapter 4). It is also a useful reminder that negotiation can (and does) continue through all the stages. It is often said that negotiation continues right to the end. Lawyers even have a term for negotiated settlement late in the day: *on the court steps*.

Stage 2: mediation, conciliation, neutral evaluation, construction adjudication

Stage 2 includes many processes: mediation, conciliation, neutral evaluation, construction adjudication and, of course, negotiation which continues throughout. Mediation has been described as assisted or facilitated negotiation, so the first of the Stage 2 techniques is an extension of the negotiation process in Stage 1.

Stage 3: arbitration, expert determination, litigation

Stage 3 might be thought of as the binding (or adjudicative) dispute resolution techniques, but of course remember that negotiation can produce settlement right up to the court steps, and arbitrators can give consent awards to formalise agreements negotiated up to and including the arbitration room.

Other authors talk of three core techniques which might be applied in the resolution of disputes. First, negotiation, which refers to the problem-solving efforts of the parties. Second, third party intervention, which does not lead to a binding decision being imposed on the parties, and finally, the adjudicative process, the ultimate outcome of which is an imposed binding decision. Mackie *et al.* describe such an approach as the 'three pillars' of dispute resolution.[1] The discrete techniques can be introduced under one of the three pillars, depending upon the main characteristics of the particular technique (see Figure 11.1).

Consideration of the stages of dispute resolution is valuable as it helps understand the techniques used, and if the classification of stages is robust it will allow meaningful analysis. The two classifications examined here

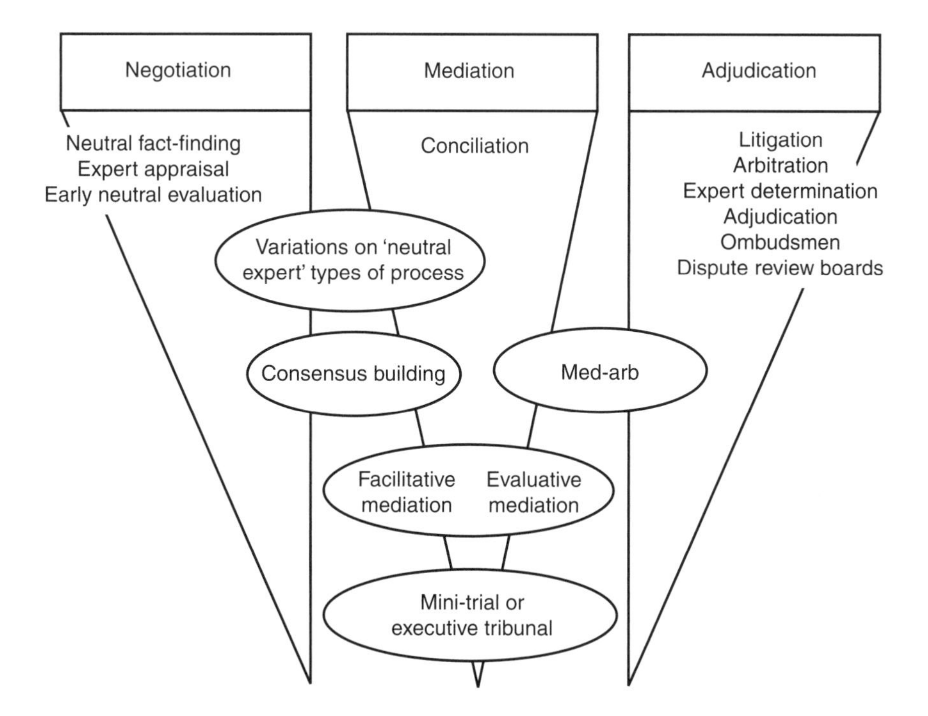

Figure 11.1 'The dispute resolution landscape'
Source: adapted from Mackie, K. Miles, D. and Marsh, W. (1995) *Commercial Dispute Resolution*, London: Butterworths.

reach similar conclusions: three stages starting with negotiation, then a third party intervention stage with a third stage of an adjudicative process. Using the classification developed in Chapters 2 and 3, it is suggested that the three-stage dispute resolution model is robust and appropriate; however, a preliminary or precursor first stage is important: conflict management and dispute avoidance. This reiterates the view that the first stage in dispute resolution is dispute avoidance. The four-stage model is then:

1 Conflict management (or dispute avoidance).
2 Negotiation as dispute resolution.
3 Non-binding (or consensual) dispute resolution.
4 Binding (or adjudicative) dispute resolution.

Conflict management (or dispute avoidance)

As pointed out in Chapter 3, conflict management should be the biggest part of any book – the avoidance of disputes is as important, if not more important, than the resolution of disputes. It is interesting that publications on dispute avoidance are dwarfed by those on dispute resolution. Planning is crucial: take any steps necessary to avoid dispute. An American term 'stovepiping' is a useful reminder that during enquiries a contractor will direct an employer to certain previous customers so that only good reviews are given. Do the 'due diligence' and do it well, choose the party that you contract with carefully. It is axiomatic that dispute avoidance costs money, but it is also axiomatic that dispute avoidance is value.

Sometimes this axiom is known as the Common Law of Business Balance:

> It's unwise to pay too much, but it's worse to pay too little. When you pay too much, you lose a little money – that is all. When you pay too little, you sometimes lose everything, because the thing you bought was incapable of doing the thing it was bought to do. The common law of business balance prohibits paying a little and getting a lot – it can't be done. If you deal with the lowest bidder, it is well to add something for the risk you run, and if you do that you will have enough to pay for something better.

This quotation has been widely attributed to John Ruskin, but has never been sourced to any of his works.

Negotiation as dispute resolution

The next best thing to avoiding disputes is using simple negotiation to resolve the dispute. Negotiation is cheap, quick and efficient. If it was that easy the next two stages wouldn't be needed. A common question is: Why do negotiations fail? Any internet search produces many results, usually containing someone's homily about the *x* most common reasons about why negotiations fail. It is reiterated, over and over again here, that negotiation can continue through all the stages, and that negotiation is the problem-solving efforts of the parties.

Non-binding (or consensual) dispute resolution

It is felt by many that negotiations fail because the parties need a helping hand. The analogy of chemical catalyst is made in Chapter 4. The next stage, then, in dispute resolution is to get the parties to reach a consensual settlement. Some mediators refer to facilitated negotiation, where a third party neutral/mediator/facilitator assists the parties to find a consensual solution.

Binding (or adjudicative) dispute resolution

There must be issues where the parties need an adjudicative decision. There may be points of law that have to be decided, there may be points of national interest. Pubic bodies might require a third party adjudicative decision. There are many reasons. Some would point to adjudicative dispute resolution as a failure. However, binding dispute resolution is a fundamental activity of government. What is essential is that conflict management and dispute resolution is done in as efficient, effective and fair a manner as possible.

Multi-tiered dispute resolution

The concept of stages in dispute resolution has been taken forward into contracts and projects. Again, construction might be seen as a trail-blazer and the examples given here are construction ones. The process, known as multi-tiered dispute resolution, involves resolving disputes via a clause which provides for separate processes at distinct and escalating stages. There may be two, three or four tiers.

- Two-tier dispute resolution
  Example: the Channel Tunnel between Britain and France. Here, disputes were, in the first instance, referred to a panel of experts, then, as a second stage, the dispute could be referred to arbitration.[2]
- Three-tier dispute resolution
  Example: FIDIC contracts. The dispute is referred to a dispute adjudication board, then amicable settlements made if either party issues a notice of dissatisfaction and finally arbitration.[3]

- Four-tier dispute resolution
  Example: Hong Kong Airport. The dispute is referred to the engineer, then mediation, then adjudication, then arbitration.[4]

This solution of providing a staged approach towards disputes in contracts has been well received. Project-based industries have a need to keep the project moving towards completion, that is 'on-track' if a dispute arises. Staged approaches employ techniques to provide for Stage 2 techniques, which provide an adjudicative decision which is binding until the project is complete.[5]

The use of multi-tiered dispute resolution is rising both domestically and internationally, reflecting an increased realisation that a contingency-based approach to disputes is efficient. That is, the technique to resolve a dispute is contingent on many things: the dispute, the stage and the solution required. Multi-tiered dispute resolution is adaptable in its nature and can accommodate these requirements.

It is possible to have no contractual dispute resolution mechanism; everything could be left to the parties' negotiation and then to an ad hoc agreement. Experience tells that there has to be some arrangement; it is difficult, if not impossible, to find any standard contract that does not provide some mechanism. Absence of mechanism generates uncertainty and no control. Once started on that route it is logical to provide a menu of alternatives which allow the parties to make best use of the techniques and procedures available to them. Multi-tiered dispute resolution allows that choice and provides for the techniques and procedures which permit efficiency.

12 The psychology of conflict management and disputes

Pink ribbon

Psychology is the science of mind and behaviour; it seeks to understand behaviour and mental processes. It is clear, then, that psychology has a big part to play in conflict management and dispute resolution. This is a book about conflict management and dispute resolution, not a book about psychology, but two psychological techniques are chosen. The two techniques: the study of body language and the prediction of lying (or at least mendacity). These two techniques show the attraction of genuine psychology to conflict management and dispute resolution and the dangers of dipping into a rigorous science and using techniques without proper understanding.

Body language is a form of non-verbal communication, in which some advocates say the study of certain signs, cues or markers, for example body posture, gestures, facial expressions and eye movements, allows general conclusions to be drawn. Humans send and interpret body language almost entirely subconsciously, and the ability to turn this into a conscious cognitive occurrence presents obvious benefits to both dispute resolution and those seeking to manage conflict.

The prediction of untruths, or the ability to spot lying, is the holy grail of some. It is an area of great controversy. Do liars exhibit specific body language at the moment they tell a lie? Can certain physical signs be detected which indicate lying? For example, polygraph tests are admissible evidence in some jurisdiction and not in others; polygraphy is widely scorned by academics. Again, the ability to 'spot' or predict lying is another area that presents obvious benefits to both dispute resolution and those seeking to manage conflict.

It is often claimed that dispute resolvers must understand the psychology of conflict and dispute. Or in their publicity material, that certain dispute resolvers understand the psychology of dispute; for example *our mediators* (replace mediator with any conflict management or dispute resolution technique) *understand the psychology of disputes*. What, then, is the psychology of conflict and dispute?

Introduction to psychology

Psychology is the scientific study of the mind, and its aim is to understand the mental and behavioural charateristics of humans and anmials. Many claim that psychology is a relatively new science, unlike physics or chemistry, and this is often used as a pejorative. Critics equally use the term 'soft science', implying that psychology is somehow inferior than the 'hard' or mature sciences. Psychologists explore many concepts, such as perception, cognition, attention, emotion, motivation, personality, behaviour and interpersonal relationship. Psychology incorporates research from many disciplines: the social sciences, natural sciences and humanities. To counter the criticisms, psychology adopts extreme rigour in its study, research and publications. Psychologists attempt to understand the role of mental functions in individual, organisational and social relationships and behaviour.

Psychology began as a philosophical study and then sat side by side in both philosophy and biology. When it stepped apart from philosophy and biology to become its own discipline it is difficult to define, but most agree that, since it became a field in its own right, it has followed schools of thought. There are many schools but an attempt to show the schools chronologically runs the danger of falling into the trap this book seeks to avoid. This is a book about conflict management and dispute resolution not a book about psychology. If you want to study the schools of psychology, then read the literature.

Structuralism and functionalism

Structuralism was the first school of psychology, and focused on breaking down mental processes into the most basic components. Wilhelm Wund is credited as the father of experimental psychology.

The functionalism school of thought formed as a reaction to the theories of the structuralist school of thought and was led by the work of William James who thought that psychology should be pragmatic and have real practical value. This practical value should concentrate on providing benefit to individuals by explaining the function of the mind and mental processes.

Gestalt

In response to the reductionist approach of structuralism, which broke things down into their component parts (sometimes called a molecular approach), the Gestalt school of thought developed. Here, a holistic view maintains that what is important is the whole experience, and the whole is different from the sum of the parts.

Psychoanalysis

Psychoanalysis is the school of thought perhaps best known popularly, founded by Sigmund Freud in the late nineteenth and early twentieth centuries. Psychoanalysis emphasises the influence of the unconscious mind on behaviour. Freud believed that the human mind was composed of three elements: the id, the ego and the super ego. This school of thought exerts considerable influence.

Behaviourist and humanistic schools of thought

In the middle years of the twentieth century, two schools of thought developed: behaviourist and humanistic. Humanistic school developed as a response to psychoanalysis and instead focused on individual free will, personal growth and self-actualisation. Maslow with his hierarchy of needs was a considerable influence. Behaviourism, based on observation, suggests that all behaviour can be explained by environmental causes and became the predominant school of thought.

Cognitive psychology

Cognitive psychology is the school of psychology that studies mental processes including how people think, perceive, remember and learn. Cognitive psychology is part of a discipline of cognitive science. This branch of psychology is related to other disciplines, including neuroscience, philosophy and linguistics.

What is current psychology?

This is a common question and the common answer is that psychologists use objective scientific methods to understand, explain and predict human behaviour. If the question for this book becomes: What can psychology add to conflict and dispute? Then, the answer becomes: use the methods of psychology to manage conflict and facilitate the effective resolution of disputes. That begs the question: What are these methods? There is a danger in dipping into a discipline to cherry-pick methods or techniques. These dangers can be highlighted by looking at two methods that many would consider as methods or techniques of psychology: body language and polygraphy.

Body language

Body language is the most common example of a wider discipline kinesics: the study of movements, including posture.

Albert Mehrabian, a psychologist who started the discipline, is credited for finding a 7–38–55 per cent rule, denoting how much communication was conferred by words, tone and body language.

Mehrabian's rule

Words	7 per cent
Tone	38 per cent
Body language	55 per cent

These are often abbreviated to the three Vs: verbal (words) seven per cent, vocal (tone) 38 per cent, and visual (body language) 55 per cent.

But of course this is an oversimplification, or, some say, a misinterpretation of what Mehrabian actually says. Nevertheless, authors go on to develop elaborate theories on the meanings of some action. Below are some examples.

- Deceit, lying or the act of withholding information can supposedly be indicated by touching the face during conversation, or blinking is an indicator of lying. Confusingly, it is also claimed that the absence of blinking can also represent lying. Eye contact is widely claimed as a predictor of deception, *he couldn't look me in the eye* is a common phrase. The problem is that there might be many reasons why he couldn't look you in the eye, he might be uncomfortable with the subject, it might not be the social mores in his culture, and so on.
- Crossing arms across chest. One of the most basic and powerful body-language signals. Supposedly this indicates that a person is putting up an unconscious barrier. It might also indicate that the person has folded their arms, or that their arms are cold. During a friendly exchange it is suggested that this indicates that a person is thinking deeply about what is being discussed. But in a conflict situation, it is suggested that it might mean that a person is expressing opposition, particularly if the person is leaning away from the speaker. A harsh or blank facial expression often indicates outright hostility.
- Consistent eye contact supposedly indicates that a person is thinking positively of what the speaker is saying. Again, confusingly, it can also mean that the other person does not trust the speaker. Does not trust them enough to take their eyes off the speaker for any length of time. On the other hand, lack of eye contact can indicate negativity. There are gender issues, with some claiming that men avoid eye contact with other men while women both need and encourage it when talking to other women.

- Mistrust or disbelief it is suggested might be indicated by an averted gaze, or by touching the ear or scratching the chin. Or they might indicate itchy ears or chins or a desire to look at something else. Deceit, lying or the act of withholding information can supposedly be indicated by touching the face during conversation, or blinking is a well-known indicator of someone who is lying. Confusingly, the absence of blinking can also represent lying as a more reliable factor than excessive blinking.
- Boredom is indicated by the head tilting to one side, or by the eyes looking straight at the speaker but becoming slightly unfocused. A head tilt may also indicate a serious medical condition or just a crick in the neck, and unfocused eyes may indicate eye problems or tiredness or drink – the list goes on.

Kinestics is a complex controversial technique. At any level the 'cues' suggested are open to interpretation. Common sense tells us to be sceptical and some counsel (this author included) that if the cues are of any use at all, they certainly do not work across cultures. However, recent thinking suggests that psychologists and anthropologists have found a way to reduce the confusing human cues to just a few categories:

- Emblems are non-verbal cues that clearly represent a verbal message, like a 'thumbs-up' gesture or the hand signal for 'okay'.
- Illustrators: some people clearly talk with their hands and the gestures used by illustrators underscore the meaning of the verbal message.
- Affect displays are facial gestures that clearly carry or convey a non-verbal message (a grimace, a smile, a frown). But beware – as ever there are cultural nuances.
- Regulators are non-verbal cues that determine the efficiency and effectiveness of the verbal communication. Body language cues that indicate the person has heard and/or understands what has been said, for example, shaking or nodding the head.

Polygraphy

A polygraph or lie detector measures several indicators while a subject is asked to answer a series of questions. The theory is that deceptive answers will produce physiological responses; that is, lying makes a human change and exhibit signals that can be detected. Then lying, deceit and untruthful behaviour can be identified by the polygraph. This theory is widely ridiculed by scientists; any internet search will support this, but polygraphs are widely used and are admitted as evidence in some jurisdictions; mostly in the USA. In 2007, polygraph testimony was admitted in 19 states as balance and counter argument. It is reported that

> in most European jurisdictions, polygraphs are not considered reliable evidence and are not generally used by police forces. Courts themselves do not order or pay for polygraph tests. In most cases, polygraph tests are voluntarily taken by a defendant in order to substantiate his or her claims.

Psychotherapy

Psychotherapy is a process that allows a person to come to a fuller understanding of their abilities, difficulties, motivations or worries in conjunction with a psychotherapist. The difference between counselling and therapy is a hotly debated issue. For the purposes of this book it is easy to exclude psychotherapy, for psychotherapists might be dispute resolvers, but rarely for commercial conflict and dispute. However, as ever, there may be valuable lessons that can be learned from psychotherapists.

The psychology of conflict and dispute

It is often claimed that dispute resolvers must understand the psychology of conflict and dispute. This claim is often without any definition of what is the psychology of conflict and dispute.

Equally, the claim is often made in dispute resolver's publicity material, that they understand the psychology of dispute. This is particularly prominent among mediators and might be given as their USP, for example, *our mediators understand the psychology of disputes.*

What, then, is the psychology of conflict and dispute? Human beings are rational beings and it is irrational to be in dispute. Economists talk of bounded rationality; the notion that in decision making, rationality of individuals is limited by the information they have, the cognitive limitations of their minds and the finite amount of time they have to make decisions. Psychologists talk of rationality wars the failure of human rationality.

Conflict management and dispute resolution might usefully use psychology methods and techniques, but the danger in non-psychologists using psychology is that they are not psychologists. They might use the wrong technique in the circumstances. Two techniques discussed here are controversial, and the dangers are obvious.

Appendix

1958 – Convention on the recognition and enforcement of foreign arbitral awards

This data is updated whenever the UNCITRAL Secretariat is informed of changes in status of the Convention.

The date format used here is DD/MM/YYYY. (Pink ribbon note: this is important because the convention in the USA is MM/DD/YYYY, which can cause confusion and embarrassment.)

State	*Notes*	*Signature*	*Ratification, Accession(*) or Succession(§)*	*Entry into force*
Afghanistan	(a), (b)		30/11/2004(*)	28/02/2005
Albania			27/06/2001(*)	25/09/2001
Algeria	(a), (b)		07/02/1989(*)	08/05/1989
Antigua and Barbuda	(a), (b)		02/02/1989(*)	03/05/1989
Argentina	(a), (b)	26/08/1958	14/03/1989	12/06/1989
Armenia	(a), (b)		29/12/1997(*)	29/03/1998
Australia			26/03/1975(*)	24/06/1975
Austria			02/05/1961(*)	31/07/1961
Azerbaijan			29/02/2000(*)	29/05/2000
Bahamas			20/12/2006(*)	20/03/2007
Bahrain	(a), (b)		06/04/1988(*)	05/07/1988
Bangladesh			06/05/1992(*)	04/08/1992
Barbados	(a), (b)		16/03/1993(*)	14/06/1993
Belarus	(e)	29/12/1958	15/11/1960	13/02/1961

State	*Notes*	*Signature*	*Ratification, Accession(*) or Succession(§)*	*Entry into force*
Belgium	(a)	10/06/1958	18/08/1975	16/11/1975
Benin			16/05/1974(*)	14/08/1974
Bolivia			28/04/1995(*)	27/07/1995
Bosnia and Herzegovina	(a), (b), (f)		01/09/1993(§)	06/03/1992
Botswana	(a), (b)		20/12/1971(*)	19/03/1972
Brazil			07/06/2002(*)	05/09/2002
Brunei Darussalam	(a)		25/07/1996(*)	23/10/1996
Bulgaria	(a), (e)	17/12/1958	10/10/1961	08/01/1962
Burkina Faso			23/03/1987(*)	21/06/1987
Cambodia			05/01/1960(*)	04/04/1960
Cameroon			19/02/1988(*)	19/05/1988
Canada	(h)		12/05/1986(*)	10/08/1986
Central African Republic	(a), (b)		15/10/1962(*)	13/01/1963
Chile			04/09/1975(*)	03/12/1975
China	(a), (b), (j)		22/01/1987(*)	22/04/1987
Colombia			25/09/1979(*)	24/12/1979
Cook Islands			12/01/2009(*)	12/04/2009
Costa Rica		10/06/1958	26/10/1987	24/01/1988
Côte d'Ivoire			01/02/1991(*)	02/05/1991
Croatia	(a), (b), (f)		26/07/1993(§)	08/10/1991
Cuba	(a), (b)		30/12/1974(*)	30/03/1975
Cyprus	(a), (b)		29/12/1980(*)	29/03/1981
Czech Republic	(a), (e)		30/09/1993(§)	01/01/1993
Denmark	(a), (b), (c)		22/12/1972(*)	22/03/1973

State	*Notes*	*Signature*	*Ratification, Accession(*) or Succession(§)*	*Entry into force*
Djibouti	(a), (b)		14/06/1983(§)	27/06/1977
Dominica			28/10/1988(*)	26/01/1989
Dominican Republic			11/04/2002(*)	10/07/2002
Ecuador	(a), (b)	17/12/1958	03/01/1962	03/04/1962
Egypt			09/03/1959(*)	07/06/1959
El Salvador		10/06/1958	26/02/1998	27/05/1998
Estonia			30/08/1993(*)	28/11/1993
Fiji			27/09/2010(*)	26/12/2010
Finland		29/12/1958	19/01/1962	19/04/1962
France	(a)	25/11/1958	26/06/1959	24/09/1959
Gabon			15/12/2006(*)	15/03/2007
Georgia			02/06/1994(*)	31/08/1994
Germany	(a)	10/06/1958	30/06/1961	28/09/1961
Ghana			09/04/1968(*)	08/07/1968
Greece	(a), (b)		16/07/1962(*)	14/10/1962
Guatemala	(a), (b)		21/03/1984(*)	19/06/1984
Guinea			23/01/1991(*)	23/04/1991
Haiti			05/12/1983(*)	04/03/1984
Holy See	(a), (b)		14/05/1975(*)	12/08/1975
Honduras			03/10/2000(*)	01/01/2001
Hungary	(a), (b)		05/03/1962(*)	03/06/1962
Iceland			24/01/2002(*)	24/04/2002
India	(a), (b)	10/06/1958	13/07/1960	11/10/1960
Indonesia	(a), (b)		07/10/1981(*)	05/01/1982
Iran	(a), (b)		15/10/2001(*)	13/01/2002

State	*Notes*	*Signature*	*Ratification, Accession(*) or Succession(§)*	*Entry into force*
Ireland	(a)		12/05/1981(*)	10/08/1981
Israel		10/06/1958	05/01/1959	07/06/1959
Italy			31/01/1969(*)	01/05/1969
Jamaica	(a), (b)		10/07/2002(*)	08/10/2002
Japan	(a)		20/06/1961(*)	18/09/1961
Jordan		10/06/1958	15/11/1979	13/02/1980
Kazakhstan			20/11/1995(*)	18/02/1996
Kenya	(a)		10/02/1989(*)	11/05/1989
Kuwait	(a)		28/04/1978(*)	27/07/1978
Kyrgyzstan			18/12/1996(*)	18/03/1997
Lao People's Democratic Republic			17/06/1998(*)	15/09/1998
Latvia			14/04/1992(*)	13/07/1992
Lebanon	(a)		11/08/1998(*)	09/11/1998
Lesotho			13/06/1989(*)	11/09/1989
Liberia			16/09/2005(*)	15/12/2005
Lithuania	(e)		14/03/1995(*)	12/06/1995
Luxembourg	(a)	11/11/1958	09/09/1983	08/12/1983
Madagascar	(a), (b)		16/07/1962(*)	14/10/1962
Malaysia	(a), (b)		05/11/1985(*)	03/02/1986
Mali			08/09/1994(*)	07/12/1994
Malta	(a), (f)		22/06/2000(*)	20/09/2000
Marshall Islands			21/12/2006(*)	21/03/2007
Mauritania			30/01/1997(*)	30/04/1997
Mauritius	(a)		19/06/1996(*)	17/09/1996
Mexico			14/04/1971(*)	13/07/1971

State	*Notes*	*Signature*	*Ratification, Accession(*) or Succession(§)*	*Entry into force*
Monaco	(a), (b)	31/12/1958	02/06/1982	31/08/1982
Mongolia	(a), (b)		24/10/1994(*)	22/01/1995
Montenegro	(a), (b), (f)		23/10/2006(§)	03/06/2006
Morocco	(a)		12/02/1959(*)	07/06/1959
Mozambique	(a)		11/06/1998(*)	09/09/1998
Nepal	(a), (b)		04/03/1998(*)	02/06/1998
Netherlands	(a), (d)	10/06/1958	24/04/1964	23/07/1964
New Zealand	(a)		06/01/1983(*)	06/04/1983
Nicaragua			24/09/2003(*)	23/12/2003
Niger			14/10/1964(*)	12/01/1965
Nigeria	(a), (b)		17/03/1970(*)	15/06/1970
Norway	(a), (i)		14/03/1961(*)	12/06/1961
Oman			25/02/1999(*)	26/05/1999
Pakistan	(a)	30/12/1958	14/07/2005	12/10/2005
Panama			10/10/1984(*)	08/01/1985
Paraguay			08/10/1997(*)	06/01/1998
Peru			07/07/1988(*)	05/10/1988
Philippines	(a), (b)	10/06/1958	06/07/1967	04/10/1967
Poland	(a), (b)	10/06/1958	03/10/1961	01/01/1962
Portugal	(a)		18/10/1994(*)	16/01/1995
Qatar			30/12/2002(*)	30/03/2003
Republic of Korea	(a), (b)		08/02/1973(*)	09/05/1973
Republic of Moldova	(a), (f)		18/09/1998(*)	17/12/1998
Romania	(a), (b), (e)		13/09/1961(*)	12/12/1961

State	*Notes*	*Signature*	*Ratification, Accession(*) or Succession(§)*	*Entry into force*
Russian Federation	(e)	29/12/1958	24/08/1960	22/11/1960
Rwanda			31/10/2008(*)	29/01/2009
Saint Vincent and the Grenadines	(a), (b)		12/09/2000(*)	11/12/2000
San Marino			17/05/1979(*)	15/08/1979
Saudi Arabia	(a)		19/04/1994(*)	18/07/1994
Senegal			17/10/1994(*)	15/01/1995
Serbia	(a), (b), (f)		12/03/2001(§)	27/04/1992
Singapore	(a)		21/08/1986(*)	19/11/1986
Slovakia	(a), (e)		28/05/1993(§)	01/01/1993
Slovenia	(f), (k)		06/07/1992(§)	25/06/1991
South Africa			03/05/1976(*)	01/08/1976
Spain			12/05/1977(*)	10/08/1977
Sri Lanka		30/12/1958	09/04/1962	08/07/1962
Sweden		23/12/1958	28/01/1972	27/04/1972
Switzerland		29/12/1958	01/06/1965	30/08/1965
Syrian Arab Republic			09/03/1959(*)	07/06/1959
Thailand			21/12/1959(*)	20/03/1960
The former Yugoslav Republic of Macedonia	(b), (f), (l)		10/03/1994(§)	17/11/1991
Trinidad and Tobago	(a), (b)		14/02/1966(*)	15/05/1966
Tunisia	(a), (b)		17/07/1967(*)	15/10/1967
Turkey	(a), (b)		02/07/1992(*)	30/09/1992
Uganda	(a)		12/02/1992(*)	12/05/1992

State	*Notes*	*Signature*	*Ratification, Accession(*) or Succession(§)*	*Entry into force*
Ukraine	(e)	29/12/1958	10/10/1960	08/01/1961
United Arab Emirates			21/08/2006(*)	19/11/2006
United Kingdom of Great Britain and Northern Ireland	(a), (g)		24/09/1975(*)	23/12/1975
United Republic of Tanzania	(a)		13/10/1964(*)	11/01/1965
United States of America	(a), (b)		30/09/1970(*)	29/12/1970
Uruguay			30/03/1983(*)	28/06/1983
Uzbekistan			07/02/1996(*)	07/05/1996
Venezuela (Bolivarian Republic of)	(a), (b)		08/02/1995(*)	09/05/1995
Viet Nam	(a), (b), (e)		12/09/1995(*)	11/12/1995
Zambia			14/03/2002(*)	12/06/2002
Zimbabwe			29/09/1994(*)	28/12/1994

Parties: 145

a Declarations and reservations. This State will apply the Convention only to recognition and enforcement of awards made in the territory of another contracting State.

b Declarations and reservations. This State will apply the Convention only to differences arising out of legal relationships, whether contractual or not, that are considered commercial under the national law.

c On 10 February 1976, Denmark declared that the Convention shall apply to the Faeroe Islands and Greenland.

d On 24 April 1964, the Netherlands declared that the Convention shall apply to the Netherlands Antilles.

e Declarations and reservations. With regard to awards made in the territory of non-contracting States, this State will apply the Convention only to the extent to which those States grant reciprocal treatment.

f Declarations and reservations. This State will apply the Convention only to those arbitral awards which were adopted after the entry into effect of the Convention.

g The United Kingdom extended the territorial application of the Convention, for the case of awards made only in the territory of another contracting State, to the following territories: Gibraltar (24 September 1975), Isle of Man (22 February 1979), Bermuda (14 November 1979), Cayman Islands (26 November 1980), Guernsey (19 April 1985), Jersey (28 May 2002).

h Declarations and reservations. Canada declared that it would apply the Convention only to differences arising out of legal relationships, whether contractual or not, that were considered commercial under the laws of Canada, except in the case of the Province of Quebec, where the law did not provide for such limitation.

i This State will not apply the Convention to differences where the subject matter of the proceedings is immovable property situated in the State, or a right in or to such property.

j Upon resumption of sovereignty over Hong Kong on 1 July 1997, the Government of China extended the territorial application of the Convention to Hong Kong, Special Administrative Region of China, subject to the statement originally made by China upon accession to the Convention. On 19 July 2005, China declared that the Convention shall apply to the Macao Special Administrative Region of China, subject to the statement originally made by China upon accession to the Convention.

k On 4 June 2008, Slovenia withdrew the declarations made upon succession mentioned in footnotes (a) and (b).

l On 16 September 2009, The former Yugoslav Republic of Macedonia withdrew the declaration made upon succession mentioned in footnote (a).

Source: http://www.uncitral.org/uncitral/en/uncitral_texts/arbitration/NYConvention_status.html.

Notes

1 Introduction

1 The Pound conference, 'Perspectives on justice in the future, proceedings of the national nonference on the causes of popular dissatisfaction with the administration of justice'. Published in 1979.

2 Conflict management and dispute resolution

1 Campell T. (1981) *Seven theories of human society*, Oxford: Oxford University Press.
2 Metcalf, H. (2003) *Dynamic Administration: The Collected Papers of Mary Parker Follett: Early Sociology of Management and Organizations*, London: Routledge.
3 Burton, J. W. (1993) 'Conflict resolution as a political philosophy', in *Conflict Resolution Theory and Practice: Integration and Application* (eds H. van der Merwe and D. J. D. Sandole), Manchester: Manchester University Press.
4 De Bono, E. (1985) *Conflicts*, London: Penguin.
5 Brown, H. and Marriot, A. (1994) *ADR Principles and Practice*. London: Sweet and Maxwell.
6 23 Con LR.
7 2 Lloyd's Rep 387.
8 See Chapter 6.
9 (2002) EWHC 2914 TCC.
10 (1998) 1 WLR 727.
11 (2003) EWHC 312.
12 (2003) EWHC 1187 TCC.
13 (2005) EWCA Civ 1358.
14 (2003) BLR 316.
15 http://www.ogc.gov.uk/documents/dispute_resolution.pdf (accessed 20th April 2011).
16 Sander, F. (1976) *The Multi-door Court House*. 70 F.R.D. 111, Harvard.
17 Lipsky, D. B. and Seeber, R. L (1998) *The Appropriate Resolution of Corporate Disputes: A Report on the Growing Use of ADR by US Corporations*. Cornell, NY: Institute on Conflict Resolution.
18 http://www.justice.gov.uk/publications/docs/alternative-dispute-resolution-08-09.pdf (accessed 20th April 2011).
19 http://www.justice.gov.uk/publications/docs/alternative-dispute-resolution-08-09.pdf (accessed 20th April 2011).

20 (2001) ADR.L.R. 12/14.
21 Rimington, S. (2002) *Open Secret: The Autobiography Of The Former Director-General Of MI5*, London: Arrow, p.228.
22 Latham, M. (1994) *Constructing the Team*, London: HMSO. National Audit Office (2001) *Modernising Construction*, London: HMSO.
23 Kumaraswamy, M. (1997) 'Conflicts, claims and disputes in construction', *Construction Law Journal,* 13, pp.21–34.
24 Department of the Environment (1996) *Making the Scheme for Construction Contracts.* A consultation paper issued by the Department of the Environment, London.
25 Latham, M. (1994) *Constructing the Team*, London: HMSO.
26 Popper, K. (2010) *The Poverty of Historicism*, London: Routledge.
27 Blimes, L. and Stiglitz, J. (2008) *The Three Trillion Dollar War: The True Cost of the Iraq Conflict*, New York: W. W. Norton & Company.
28 http://www.judiciary.gov.uk/Resources/JCO/Documents/jackson-final-report-140110.pdf (accessed 23rd October 2010).
29 Woolf, Lord (1995) *Access to Justice*, Interim Report to the Lord Chancellor on the civil justice system in England and Wales, London: HMSO.
30 Fenn, P. and Black, M. (1999) 'A survey of domestic construction arbitration in the UK', *Arbitration*, 65(3), pp.217–28.
31 http://www.statistics.gov.uk/downloads/theme_commerce/CSA-2009/Opening-page.pdf (accessed 20th April 2011).
32 Menor, L. Ramasastry, C. (2004) *Dabbawallahs of Mumbai.* HBR case study, publication date: 26 April 2004. Prod. #: 904D11-PDF-ENG.

3 Conflict management and dispute avoidance

1 Further reading: Doherty, S. (2008), *Heathrow's Terminal 5: History in the making*, Chichester: Wiley.
2 Brass, R. (2008) 'Flying in formation', *Supply Management*, 13 March, see http://www.supplymanagement.com/analysis/features/2008/flying-in-formation/?locale=en (accessed 6th May 2011).
3 Fullalove, S. (2004) 'NEC helps BAA deliver Heathrow T5', *NEC Users' group Newsletter*, Issue 30, August 2004, see http://www.neccontract.com/news/article.asp?NEWS_ID=512 (accessed 14th July 2011).

4 Negotiation

1 Fisher, R. and Ury, W. (1981) *Getting to Yes*, New York: Random House.
2 Ury, W. L. (1991) *Getting Past No*, New York: Random House.
3 DeBono, E. (1991) *Thinking Course: Facts on File*, New York: Infobase Publishing.

5 Mediation

1 Genn, H. *et al.* (2007) *Twisting Arms: Court Referred And Court Linked Mediation Under Judicial Pressure*, available at http://www.justice.gov.uk/publications/docs/Twisting-arms-mediation-report-Genn-et-al.pdf.
2 http://eur-lex.europa.eu/LexUriServ/LexUriServ.do?uri=OJ:L:2008:136:0003:0008:EN:PDF.
3 *Cowl and Others* v. *Plymouth City Council*, Time Law Reports, January 8 2002.
4 *Dunnett* v. *Railtrack* (2002) EWCA Civ 302.
5 Halsey and Milton Keynes General NHS Trust (2004) EWCA 3006 Civ 576.

6 *Leicester Circuits Ltd* v. *Coates Brothers plc* and *Royal Bank of Canada Trust Corporation Ltd* v. *Secretary of State for Defence* (2003), EWHC 1479 (Ch).
7 *Rolf* v. *De Guerin* (2011) EWCA Civ 78.
8 The Annual Pledge Report 2008/09: Monitoring the Effectiveness of the Government's commitment to using Alternative Dispute Resolution, The Ministry of Justice.
9 http://eur-lex.europa.eu/LexUriServ/LexUriServ.do?uri=OJ:L:2008:136:0003:0008:EN:PDF.

6 Construction adjudication

1 Bentley, B. (1992) *Adjudication Procedures: A Temporary Diversion*, in P. Fenn, and R. Gameson (eds), *Construction Conflict Management and Resolution*, London: EF & N Spon.
2 *Channel Group* v. *Balfour Beatty Ltd.* (1993) A.C. 347.
3 *Surveyors Acting as Adjudicators in the Construction Industry: A Guide to Best Practice*, 2nd edition, RICS.
4 Fenn P. and O'Shea M. (2008) 'Adjudication: tiered and temporary binding dispute resolution', in *Construction and Engineering, Journal of Professional Issues in Engineering Education and Practice*, 134(2), p.203, DOI:10.1061/(ASCE)1052-3928(2008)134:2(203). Invited editorial.

7 Arbitration

1 Stephenson, D. (1987) *Arbitration for Contractors*, London: Construction News Books.
2 Section 10 Arbitration Act 1950, Section 18 Arbitration Act 1996.
3 http://abta.com/consumer-services/travel_problems/arbitration.
4 http://www.travelredress.co.uk/?p=4&lang=e.
5 Mustill, M. and Boyd, S. (1999) *Commercial Arbitration*, London: Butterworths.
6 The leading Scholar is Prof Derek Roebuck, e.g. Roebuck, D., deLoynes and de Fumichon (2004) *Roman Arbitration*, London: Holo Books.
7 See for example http://www.icj-cij.org/court/index.php?p1=1&p2=1.
8 Crowter, H. (1998) *Introduction to Arbitration*, London: LLP.
9 http://www.uncitral.org/uncitral/en/index.html.
10 http://www.uncitral.org/uncitral/en/uncitral_texts/arbitration.html.
11 http://www.legislation.gov.uk/ukpga/1996/23/contents.
12 http://www.legislation.gov.uk/ukpga/1996/23/contents.
13 http://www.uncitral.org/uncitral/en/uncitral_texts/arbitration/NYConvention.html.
14 Moses, M. (2008), *The Principles and Practice of International Commercial Arbitration*, Cambridge: Cambridge University Press.
15 http://www.pwc.co.uk/eng/publications/international_arbitration_2008.html.

8 Other ADR techniques

1 Fenn, P. and Gameson, R. (1992). *Construction Conflict Management and Resolution*, London: EF & N Spon.
2 Sanchez, V. (1996) 'Towards a history of ADR: the dispute processing continuum in Anglo-Saxon England and today', *The Ohio State Journal on Dispute Resolution*, 11(1), pp.1–39.
3 Gould, N., Capper, P., Dixon, G. and Cohen, M. (1999) *Dispute Resolution in the Construction Industry*, London: Thomas Telford Publishing.

4 This section draws heavily on a paper to the Society of Construction Law 4 May 1999. HHJ J Toulmin and Robert Stevenson, Early Neutral Evaluation in the Technology and Construction Court.
5 Mr Justice Cresswell, *The Times*, 10 December 1993.
6 Mr Justice Walker, *The Times*, 7 June 1996.
7 Genn, H. (1999) The Central London Count Court Pilot Mediation Scheme, LDCD 9/58.
8 See generally Kendall, J. (1996) *Dispute Resolution: Expert Determination* (2nd edition).
9 (1992) 1 W.L.R. 277.
10 (1991) 28 EG 86, itself approved by the Court of Appeal in *The 'Glacier Bay'* (1996) 1 Lloyds Rep 370, 377 col. 2.
11 (1996) 1 WLR 48.
12 (1997) 1 Lloyd's Rep 106.
13 Kendall, J. (2008) *Dispute Resolution: Expert Determination* (4th edition), London: Sweet and Maxwell.
14 *Owen Pell Ltd* v. *Bindi (London) Ltd*, Court of Appeal – Technology and Construction Court, 19 May 2008, (2008) EWHC 1420.
15 Stopping disputes before they start Commercial Lawyer Special Report, February 2001.
16 www.drb.org.
17 Wall, C. (1992) 'The Dispute Resolution Adviser in the Construction Industry', in Fenn, P. and Gameson, R. (eds) (1992) *Construction Conflict: Management and Resolution*, London: Chapman Hall.

9 Cooperation and collaboration

1 Yeung *et al.* (2007), The definition of alliancing in construction as a Wittgenstein family-resemblance concept, *International Journal of Project Management*, 25, pp.219–31.
2 Relationship Contracting: Optimising Project Outcomes, The Australian Constructors Association, http://www.constructors.com.au/publications/rc_general/Relationship%20Contracting%20Optimising%20Project%20Outcomes.pdf.
3 Yates and Duran (2007) 'Utilizing Dispute Review Boards in Relational Contracting: A Case Study', J. Profl. Issues in Engrg. Educ. and Pract. 132, 334.
4 Nalebuff and Brandenburger (2002), *Co-opetition*, London: Profile Books.
5 Axelrod, R. (1990) *The Evolution of Cooperation*, London: Penguin, p.5.
6 Ashworth, A. (1980) *Trench Warfare 1914–1918: The Live and Let Live System*, New York: Holmes and Meier.
7 Dugdale, G. (1932) *Langemarck and Cumbrai*, Shrewsbury: Wilding and Son.
8 Belton Cobb, G. (1916) *Stand to Arms*, London: Darton and Co.
9 King, P. (1938) *The War the Infantry Knew*, London: PS King.
10 Morgan, J. (1916) *Leaves from a Field Notebook*, London: MacMillan.
11 Hay, I. (1916) *The First Hundred Thousand*, London: Blackwood.

10 Game theory

1 *The Art of War* is an ancient Chinese military treatise that is attributed to Sun Tzu.
2 Brandenburger, M. and Nalebuff, J. (2002) *Co-opetition*, London: Profile Books, p.3.
3 Dixit, A. and Nalebuff, B. (2008) *The Art of Strategy*, New York: Norton.

4 Brandenburger, M. and Nalebuff, J. (2002) *Co-opetition*, London: Profile Books, p.3.
5 Von Neumann, J. and Morgenstern, O. (1944) Theory of Games and Economic Behavior, Princeton, NJ: Princeton University Press.
6 Nasar, S. (1998) *A Beautiful Mind*, New York: Simon and Schuster.
7 Giles, L. (2008) *The Art of War by Sun Tzu*, El Paso Norte Press, 2007, Wing R., *The Art of Strategy*, Main Street Books, Washington, Chow-Hou Wee (2003). Sun Zu Art of War: *An Illustrated Translation with Asian Perspectives and Insights*. Pearson Education Asia Pte Ltd.
8 See for example McNeilly, M. (1996) *Sun Tzu and the Art of Business: Six Strategic Principles for Managers*, New York: Oxford University Press.
9 Kammerer, P. (2006) 'The Art of Negotiation', *South China Morning Post*, 21 April 2006.
10 Schelling, T. (1980) *The Strategy of Conflict*, Harvard University Press.
11 Baird, Gertner and Picker (1998) *Game Theory and Law*, Harvard University Press.

11 The stages of dispute resolution

1 Mackie, K. Miles, D. and Marsh, W. (1995) *Commercial Dispute Resolution*, London: Butterworths.
2 Jones, D. (2009) 'Dealing with multi-tiered dispute resolution process', *The International Journal of Arbitration, Mediation and Dispute Management*, 75(2).
3 Federation Internationale des Ingenieur-Conseils contracts often referred to as the Rainbow Suite.
4 Connerty, A. (1996) 'The role of ADR in the resolution of international disputes', *Arbitration International*, 12(1).
5 Baker, E. (2009) *Is it all necessary? Who benefits? Provision for multi-tier dispute resolution in international construction contracts,* Society of Construction Law Paper 154.

Index

Tables have *Table* after the locator.
Figures have *Fig* after the locator.